Frontispiece: Max Planck and Albert Einstein, the originators of the two cornerstones of modern physics, quantum theory and relativity. Much effort has been devoted over the years to the synthesis of these two theories, but the quantization of the General Theory of Relativity is still incomplete. (Courtesy of the American Institute of Physics.)

PHYSICS THROUGH THE 1990s

Gravitation, Cosmology, and Cosmic-Ray Physics

Panel on Gravitation, Cosmology, and Cosmic-Ray Physics

Physics Survey Committee

Board on Physics and Astronomy

Commission on Physical Sciences, Mathematics, and Resources

National Research Council

NATIONAL ACADEMY PRESS
Washington, D.C.   1986

NATIONAL ACADEMY PRESS 2101 Constitution Avenue, NW Washington, DC 20418

NOTICE: The project that is the subject of this report was approved by the Governing Board of the National Research Council, whose members are drawn from the councils of the National Academy of Sciences, the National Academy of Engineering, and the Institute of Medicine. The members of the committee responsible for the report were chosen for their special competences and with regard for appropriate balance.

This report has been reviewed by a group other than the authors according to procedures approved by a Report Review Committee consisting of members of the National Academy of Sciences, the National Academy of Engineering, and the Institute of Medicine.

The National Research Council was established by the National Academy of Sciences in 1916 to associate the broad community of science and technology with the Academy's purposes of furthering knowledge and of advising the federal government. The Council operates in accordance with general policies determined by the Academy under the authority of its congressional charter of 1863, which establishes the Academy as a private, nonprofit, self- governing membership corporation. The Council has become the principal operating agency of both the National Academy of Sciences and the National Academy of Engineering in the conduct of their services to the government, the public, and the scientific and engineering communities. It is administered jointly by both Academies and the Institute of Medicine. The National Academy of Engineering and the Institute of Medicine were established in 1964 and 1970, respectively, under the charter of the National Academy of Sciences.

The Board on Physics and Astronomy is pleased to acknowledge generous support for the Physics Survey from the Department of Energy, the National Science Foundation, the Department of Defense, the National Aeronautics and Space Administration, the Department of Commerce, the American Physical Society, Coherent (Laser Products Division), General Electric Company, General Motors Foundation, and International Business Machines Corporation.

Library of Congress Cataloging-in-Publication Data
Main entry under title:

Gravitation, cosmology, and cosmic-ray physics.

(Physics through the 1990s)
Includes index.
1. Gravitation. 2. Cosmology. 3. Cosmic rays.
I. National Research Council (U.S.). Panel on
Gravitation, Cosmology, and Cosmic-Ray Physics.
II. Series.
QC178.G64   1986   531'.14   85-32019
ISBN 0-309-03579-1

Printed in the United States of America

PANEL ON GRAVITATION, COSMOLOGY, AND COSMIC-RAY PHYSICS

DAVID T. WILKINSON, Princeton University, *Chairman*
PETER L. BENDER, University of Colorado
DOUGLAS M. EARDLEY, University of California, Santa Barbara
THOMAS K. GAISSER, University of Delaware
JAMES B. HARTLE, University of California, Santa Barbara
MARTIN H. ISRAEL, Washington University
LAWRENCE W. JONES, University of Michigan
R. BRUCE PARTRIDGE, Haverford College
DAVID N. SCHRAMM, The University of Chicago
IRWIN I. SHAPIRO, Harvard-Smithsonian Center for Astrophysics
ROBERT F. C. VESSOT, Harvard-Smithsonian Center for Astrophysics
ROBERT V. WAGONER, Stanford University

PHYSICS SURVEY COMMITTEE

WILLIAM F. BRINKMAN, Sandia National Laboratories, *Chairman*
JOSEPH CERNY, University of California, Berkeley, and Lawrence Berkeley Laboratory
RONALD C. DAVIDSON, Massachusetts Institute of Technology
JOHN M. DAWSON, University of California, Los Angeles
MILDRED S. DRESSELHAUS, Massachusetts Institute of Technology
VAL L. FITCH, Princeton University
PAUL A. FLEURY, AT&T Bell Laboratories
WILLIAM A. FOWLER, W. K. Kellogg Radiation Laboratory
THEODOR W. HÄNSCH, Stanford University
VINCENT JACCARINO, University of California, Santa Barbara
DANIEL KLEPPNER, Massachusetts Institute of Technology
ALEXEI A. MARADUDIN, University of California, Irvine
PETER D. MACD. PARKER, Yale University
MARTIN L. PERL, Stanford University
WATT W. WEBB, Cornell University
DAVID T. WILKINSON, Princeton University

DONALD C. SHAPERO, *Staff Director*
ROBERT L. RIEMER, *Staff Officer*
CHARLES K. REED, *Consultant*

BOARD ON PHYSICS AND ASTRONOMY

HANS FRAUENFELDER, University of Illinois, *Chairman*
FELIX H. BOEHM, California Institute of Technology
RICHARD G. BREWER, IBM San Jose Research Laboratory
DEAN E. EASTMAN, IBM T.J. Watson Research Center
JAMES E. GUNN, Princeton University
LEO P. KADANOFF, The University of Chicago
W. CARL LINEBERGER, University of Colorado
NORMAN F. RAMSEY, Harvard University
MORTON S. ROBERTS, National Radio Astronomy Observatory
MARSHALL N. ROSENBLUTH, University of Texas at Austin
WILLIAM P. SLICHTER, AT&T Bell Laboratories
SAM B. TREIMAN, Princeton University

DONALD C. SHAPERO, *Staff Director*
ROBERT L. RIEMER, *Staff Officer*
HELENE PATTERSON, *Staff Assistant*
SUSAN WYATT, *Staff Assistant*

COMMISSION ON PHYSICAL SCIENCES, MATHEMATICS, AND RESOURCES

HERBERT FRIEDMAN, National Research Council, *Chairman*
THOMAS D. BARROW, Standard Oil Company (Retired)
ELKAN R. BLOUT, Harvard Medical School
WILLIAM BROWDER, Princeton University
BERNARD F. BURKE, Massachusetts Institute of Technology
GEORGE F. CARRIER, Harvard University
CHARLES L. DRAKE, Dartmouth College
MILDRED S. DRESSELHAUS, Massachusetts Institute of Technology
JOSEPH L. FISHER, Office of the Governor, Commonwealth of Virginia
JAMES C. FLETCHER, University of Pittsburgh
WILLIAM A. FOWLER, California Institute of Technology
GERHART FRIEDLANDER, Brookhaven National Laboratory
EDWARD D. GOLDBERG, Scripps Institution of Oceanography
MARY L. GOOD, Signal Research Center
J. ROSS MACDONALD, University of North Carolina
THOMAS F. MALONE, Saint Joseph College
CHARLES J. MANKIN, Oklahoma Geological Survey
PERRY L. MCCARTY, Stanford University
WILLIAM D. PHILLIPS, Mallinckrodt, Inc.
ROBERT E. SIEVERS, University of Colorado
JOHN D. SPENGLER, Harvard School of Public Health
GEORGE W. WETHERILL, Carnegie Institution of Washington

RAPHAEL G. KASPER, *Executive Director*
LAWRENCE E. MCCRAY, *Associate Executive Director*

Preface

Gravitation, cosmology, and cosmic-ray physics are often regarded as subfields of astrophysics, as well as physics, because they are practiced by using physical techniques in an astronomical setting. However, this report makes no pretense of surveying all of astrophysics; that enormous task was excellently done by the Astronomy Survey Committee (George B. Field, chairman). Their report, *Astronomy and Astrophysics for the 1980's* (National Academy Press, Washington, D.C., 1982), has been widely circulated, and its recommendations are currently being considered and implemented. We have restricted our review to the above-named three areas of physics and astrophysics currently of particular interest to physicists.

Gravitation was explicitly not considered in the Field report and thus becomes a focus of this report. Cosmology has been an active area of astronomy for 60 years, and the many successes and opportunities of astronomical techniques are eloquently described in the Field report. The cosmology part of this report attempts to supplement the report of the Astronomy Survey Committee by emphasizing new results and ideas, particularly those triggered by recent contributions from other areas of physics. There is also some overlap between this report and the Field report in the area of cosmic rays; however, the vast scope of the earlier report allowed only cursory treatment. The study of cosmic rays, developed and practiced mainly by physicists, is an appropriate topic for the present report. Choosing which areas of astrophysics not

viii   PREFACE

to emphasize in this study was more difficult. Related areas that could logically have been included are x-ray and gamma-ray astrophysics, most topics in theoretical astrophysics, nuclear astrophysics, solar physics, atomic and molecular astrophysics, and astrophysical plasmas. The interconnectedness of astrophysics leads to some discussion in our report of all of these active areas. Also, reviews and recommendations concerning some of these areas can be found in the Astronomy Survey Committee report and in the reports of other panels of the Physics Survey Committee.

In this report we have tried to characterize the fields by reporting some recent successes (Highlights) and by discussing some open questions that are guiding current research (Opportunities). The level and style of the presentation were chosen assuming that the reader is a student or a colleague not currently active in these fields. Experts will no doubt find regrettable omissions and technical errors; we did put clarity and perspective above completeness and detailed accuracy when it seemed that a choice was necessary. Our hardest task, however, was to attempt to look into the future and chart a reasonable course (Recommendations). At best one can extrapolate ahead the most promising current research and ideas, hoping that work on this predictable program will best facilitate discoveries and new directions. Indeed, we wish to emphasize that all three of these research areas are developing rapidly and that flexibility will be needed to respond effectively to new ideas and discoveries. We expect that some of our recommendations will appear quite foolish 10 years from now because of unanticipated new developments.

Our activities began with the formation of the panel in September 1983. In October about 90 "Dear Colleague" letters solicited advice from physicists and astronomers active in gravitation and cosmology. The letters requested views on facilities or major instrumentation needs, promising new areas, and a draft outline of this report. Based on that advice a meeting was called in December to consider proposed initiatives in gravitation. A list of participants and the agenda were widely circulated before the meeting. No panel meetings were held in cosmology or cosmic rays as responses to our solicitations did not indicate that meetings were needed. In these areas we relied on letters from colleagues and the comments, criticism, and advice of readers. We are particularly indebted to an active group of expert, critical readers. Their extensive comments on our first draft and guidance on the recommendations have substantially affected the content and conclusions of this report. We thank the readers: Marc Davis, University of California, Berkeley; Stanley Deser, Brandeis University;

Francis Everitt, Stanford University; George Field, Center for Astrophysics; Alan Guth, Massachusetts Institute of Technology; Peter Michelson, Stanford University; Ezra T. Newman, University of Pittsburgh; James Peebles, Princeton Universiy; Jean-Paul Richard, University of Maryland; Joseph Silk, University of California, Berkeley; Joseph Taylor, Princeton University; Kip Thorne, California Institute of Technology; V. K. Balasubrahmanyan, Goddard Spaceflight Center; Rainer Weiss, Massachusetts Institute of Technology; Clifford Will, Washington University; and Gaurang B. Yodh, University of Maryland.

The gravitation part of this report benefits greatly from the earlier report of the Space Science Board's Committee on Gravitational Physics (Irwin I. Shapiro, chairman): *Strategy for Space Research in Gravitational Physics in the 1980's*. Also, the authors of the cosmic-ray portion of this report (Thomas Gaisser, Martin Israel, and Lawrence Jones) acknowledge the assistance of the reports of NASA's Cosmic-Ray Program Working Group (1982, 1985).

The Panel is indebted to Donald C. Shapero for providing advice and services throughout this project and to Robert L. Riemer for overseeing publication of the report. Finally, we acknowledge the assistance and patience of Marion Fugill (Princeton), who held us together and made order out of the chaos of many drafts of this report.

Contents

I SUMMARY OF PRINCIPAL RECOMMENDATIONS
Recommendations on Gravitational Physics, 3
  Space Program in Gravitation, 3
  Ground-Based Studies in Gravitation, 4
  Gravitation Theory, 4
Recommendations on Cosmology, 5
  Space Program in Cosmology, 5
  Ground-Based Studies in Cosmology, 5
  Growth in Cosmology Research, 6
Recommendations on Cosmic-Ray Physics, 6
  Space Program in Cosmic Rays, 6
  Ground-Based Cosmic-Ray Studies, 7

II GRAVITATION

1 EXPERIMENTAL TESTS OF GENERAL RELATIVITY: INTRODUCTION 11

2 EXPERIMENTAL TESTS OF GENERAL RELATIVITY: HIGHLIGHTS 15
  Equivalence Principle, Eötvös to Lunar Laser Ranging, 15

Gravitational Redshift, Mössbauer to Rocketborne Maser, 17
Light Deflection, Eclipses to Radio Interferometry, 19
Signal Retardation, Newest and Most Accurate Test, 19
Perihelion Advance, Einstein's only Handle, 21
Changing Gravitational Constant, Solar-System Time Versus Atomic Time, 21
Laboratory Testing of Gravitation, Searching for the Unexpected, 22

3 EXPERIMENTAL TESTS OF GENERAL RELATIVITY: OPPORTUNITIES 24
Tests for "Magnetic" Gravitational Effects, 24
   Relativity Gyroscope Experiment, 24
   Black-Hole Jets, 26
Ranging to the Moon and Inner Planets, 27
   Radar Ranging, 28
   Ranging to Planetary Landers and Orbiters, 28
   Lunar Laser Ranging, 30
Measurement of Second-Order Solar-System Effects, 31
Gravitational Quadrupole Moment of the Sun, 33
Systems of Compact Stars, 34

4 SEARCH FOR GRAVITATIONAL WAVES: INTRODUCTION 36
Theory, 37
Sources, 38
Detectors, 40

5 SEARCH FOR GRAVITATIONAL WAVES: HIGHLIGHTS. 42
Binary Pulsar, 42
Bar Detectors, 43
Interferometric Detectors, 44
Pulsar Timing and Millisecond Pulsars, 46

    Sources of Gravitational Waves—Recent
        Developments, 47

**6 SEARCH FOR GRAVITATIONAL WAVES:
OPPORTUNITIES** 49
    Laser Interferometer Detector with 5-Kilometer
        Baseline, 49
    Bar Detector Sensitivity and Bandwidth, 52
    Observations with Bar Detectors, 54
    Pulsar Searches, 55
    Spacecraft Tracking, 55
    Space Interferometers, 56
    Event Rates and Source Calculations, 57
    Computation, 58

7 GRAVITATION THEORY: INTRODUCTION . . . 59

8 GRAVITATION THEORY: HIGHLIGHTS 61
    Neutron Stars, 61
    Gravitational Collapse and Black Holes, 62
    Quantum Particle Creation by Black Holes, 64
    Quantum Effects in the Early Universe, 64
    Alternative Theories, 65
    Exact Solutions of the Einstein Equations, 65
    Asymptotic Properties of Space-Time, 66
    Numerical Relativity, 67
    Emission of Gravitational Radiation, 67
    The Positive Energy Theorem, 68
    Quantum Field Theory in Curved Space-Time, 69
    Quantum Gravity, 69
        Supergravity, 71
        Kaluza-Klein Theories, 71

9 GRAVITATION THEORY: OPPORTUNITIES . . . 72
    Classical Gravitation, Singularities, Asymptotic
        Structure, 72
    Quantum Gravity, 73
    Astrophysical Properties of Neutron Stars and
        Black Holes, 75

Computation, 77
New Kinds of Experimental Tests, 77
Communication with Other Subfields: Gravitation Experiment, Astronomy and Astrophysics, Field Theory and Elementary-Particle Physics, Pure Mathematics, 78

10 RECOMMENDATIONS 80
Space Techniques, 80
Ground-Based Techniques, 81
Gravitation Theory, 81

III COSMOLOGY

11 INTRODUCTION—THE STANDARD MODEL . . 87

12 HIGHLIGHTS. 90
Big-Bang Nucleosynthesis, 90
Large-Scale Properties of the Universe, 92
Structure in the Universe, 94
Invisible Mass, 96
Cosmology and Grand Unification, 98
The Inflationary Universe, 99
Gravitational Lenses, 100

13 OPPORTUNITIES 101
Observations from Space, 101
Continued Ground-Based Observations, 104
Particle Physics and Cosmology, 106
Theory, 106

14 RECOMMENDATIONS. 108
Space Program, 108
Ground-Based Program, 109
Human and Computational Resources, 109

IV COSMIC RAYS

15 OVERVIEW 115

16 HIGHLIGHTS 121
    Nucleosynthesis, 123
        Isotope Ratios, 125
        Abundances of Heavy Elements, 125
        Solar Neutrinos, 126
    Acceleration, 127
        Shock Acceleration, 128
        Acceleration Fractionation, 129
        Termination of Acceleration Mechanism, 130
        High-Energy Gamma Rays, 131
        Anomalous Component, 132
    Galactic Cosmic-Ray Transport and the Interstellar Medium, 132
        Energy Dependence of Escape from Galaxy, 133
        Correlation Between Anisotropy and Energy, 135
        Secondaries from Light Nuclei, 135
        Propagation in Galactic Halo, 136
        Connection with Gamma and Radio Astronomy, 137
    High-Energy Nuclear and Particle Physics, 137
        Nucleon Decay Experiments as Cosmic-Ray Detectors, 138
        Nucleus-Nucleus Collisions, 139
        Cross Sections, Spectra, Anisotropies, and Composition of Primary Cosmic Rays Above 10^{17} Electron Volts, 140
        Magnetic Monopoles, 141

17 OPPORTUNITIES 143
    Spaceborne Experiments, 143
        Isotopes, 144
            Galactic Cosmic-Ray Isotopes, 144; Solar-Flare Isotopes, 145
        Ultraheavy Elements, 145
        High-Energy Composition and Spectra, 146
        Positrons, Antiprotons, Deuterium, and ^{3}He, 147
        Antimatter, 148
        Nucleus-Nucleus Interactions, 148
        Solar Modulation of Cosmic Rays, 149

Ground-Based Experiments, 149
  Gamma-Ray Astronomy, 149
  Air-Shower Detectors, 150
  Neutrino Astronomy, 151
  Magnetic Monopoles, 153
  Nucleon Decay Detectors, 153
  Solar Neutrinos, 154
  Future Opportunities, 155
Theory, 155

18 RECOMMENDATIONS 157
Spaceborne Experiments, 157
  Major New Programs, 158
  Continuing Programs, 159
  Studies for the Future, 160
Ground-Based Experiments, 161
  Gamma-Ray Astronomy, 162
  Highest-Energy Cosmic Rays and Extensive Air Showers, 162
  High-Energy Neutrino Astronomy, 163
  Magnetic Monopoles, 163
  Large Underground Detectors, 163
  Solar Neutrinos, 164
Theory, 164

INDEX 165

I

Summary of Principal Recommendations

This brief section summarizes the findings and principal recommendations of this report for each of the fields studied. The basis of the recommendations is solely scientific merit. We asked: what are currently the most important questions, and the most promising ways to get answers? Cost considerations played a major role only when comparing various approaches to a single scientific question.

Recommendations such as these tend to focus on large new facilities and to understate the importance of ongoing research by individuals and small groups. It is important to keep in mind that the ideas and basic research of small groups constitute the core of physics research in this country—a highly successful enterprise. Indeed, only out of these studies grow the initiatives and needs for large facilities. We wish to emphasize that U.S. research in each of the fields surveyed in this report is of high caliber. In implementing any of these recommendations care should be taken that productive ongoing work remains healthy.

Additional recommendations appear at the end of Parts II, III, and IV of this report. The scientific perspective and justification for these recommendations are presented in the sections titled Highlights and Opportunities.

RECOMMENDATIONS ON GRAVITATIONAL PHYSICS

Space Program in Gravitation

In the last two decades gravitation has evolved from a predominantly theoretical subject to a state where experimental work is making substantial contributions. Several effects predicted by general relativity have been checked experimentally and found to agree with theory to better than 1 percent accuracy. Also, basic assumptions such as the metric nature of gravity and the equivalence principle have been tested experimentally with high accuracy. Much of this rapid experimental progress is due to the careful application of space techniques to precision solar-system measurements; we are fortunate that the National Aeronautics and Space Administration (NASA) has recognized its special capabilities for experimental gravitation research. Noting that much fundamental work still remains, we recommend that NASA pursue a vigorous gravitational-physics program in the years ahead in order to maintain U.S. leadership in this fundamental area of physics.

- Test for "magnetic" gravitation
    Relativity gyroscope experiment (Gravity Probe B)
- Improve solar-system tests
    Improve laser and radar ranging to the Moon and planets
    Improve accuracy of ranging to future planetary spacecraft
- Study ideas at frontiers
    Millihertz gravity waves and second-order tests

Ground-Based Studies in Gravitation

Most ground-based research in gravitation is focused on the detection of gravitational waves. These difficult experiments are driven by the need to test a basic prediction of general relativity and by the hope to one day have an entirely new technique for exploring fundamental processes such as gravitational collapse. The National Science Foundation (NSF) has played an important role in fostering this work and is currently considering a major initiative—a Long-Baseline Gravitational-Wave Facility. We have studied this idea and enthusiastically endorse it, assuming that other ongoing work of high quality will not be adversely affected. We recommend that the NSF enhance its leadership in gravitation research by funding the Long-Baseline Facility, while continuing to support a vigorous program to search for gravitational waves with resonant bar detectors.

- Extend the search for gravity waves
    Build 5-km-baseline interferometers (10 Hz to 10 kHz)
    Improve resonant bars

Gravitation Theory

Theory plays a uniquely important role in gravitation. By exploring a wide range of theoretical possibilities it guides the field, pointing experimenters to the key questions. Currently, fundamental questions are being asked with important connections with other areas of physics and with mathematics. We urge that a healthy level of activity be fostered in this essential part of gravitation research.

- Maintain and strengthen a healthy, productive program
- Foster natural links to other areas of physics and to pure mathematics

RECOMMENDATIONS ON COSMOLOGY

Space Program in Cosmology

We are in a period of great excitement for cosmology. Our understanding of the physics of diverse cosmological epochs and processes is undergoing fundamental changes, and our meager data base is growing rapidly. Much of this growth is traceable to the highly successful U.S. space program. Besides providing unique observations from satellites, space-inspired technology has greatly enhanced the capabilities of ground-based telescopes. Looking ahead cosmologists can anticipate a decade of fascinating new data from a wide spectral range. We endorse NASA's forward-looking program and hope that the following missions of great importance to cosmology can be started soon.

- Space initiatives important to cosmology
    Advanced X-Ray Astrophysics Facility, Space Infrared Telescope Facility, Large Deployable Reflector

Ground-Based Studies in Cosmology

Astronomical telescopes have told us most of what we know about the universe, and cosmology has much to gain from the major ground-based instruments recommended by the Astronomy Survey Committee.* They will provide extreme resolution (the Very Long Baseline Array) and a much deeper view into the visible universe (the National New Technology Telescope). Recent applications of particle-physics theory to cosmology make the Superconducting Super Collider (recommended in the report of the Panel on Elementary-Particle Physics) of great interest as a probe of physics in the early universe. We wish to take note of the importance of these facilities to cosmology.

---

* Astronomy Survey Committee, National Research Council (G. B. Field, chairman) *Astronomy and Astrophysics for the 1980's* (National Academy Press, Washington, D.C., 1982).

- Major ground-based facilities important to cosmology
   Very Long Baseline Array
   National New Technology Telescope
   Superconducting Super Collider
- Maintain high quality of U.S. astronomy and astrophysics

Growth in Cosmology Research

As a rapidly growing field, drawing on many areas of physics and astronomy, cosmology has outstripped its scattered funding base. The multidisciplinary character of the field needs to be recognized and fostered. We urge the NSF to find ways to address these problems.

- Restructure support
   New funding for growing opportunities in cosmology
   Foster groups with diverse expertise

RECOMMENDATIONS ON COSMIC-RAY PHYSICS

Space Program in Cosmic Rays

Galactic cosmic rays provide a direct sample of material from outside the solar system, while solar energetic particles provide a sample of material from the Sun and the low-energy anomalous component of cosmic rays probably provides a sample of the local interstellar medium. All these energetic particles are evidence of processes in nature that accelerate particles to relativistic energies. We recommend that NASA continue a vigorous program of extended cosmic-ray observations in space in order to measure the elemental and isotopic composition of cosmic rays over a wide range of energies; measure electrons, positrons, and antiprotons; and search for heavier antimatter. These observations will address questions of nucleosynthesis and galactic chemical evolution, astrophysical particle acceleration, and the particle/antiparticle asymmetry of the universe.

- Particle Astrophysics Magnet Facility
   Superconducting magnetic spectrometer on the Space Station
- Cosmic Ray Explorer
   Spacecraft outside the magnetosphere measuring low-energy galactic cosmic rays, solar energetic particles, and anomalous cosmic rays

Ground-Based Cosmic-Ray Studies

The search for the origin of high-energy cosmic rays has long been a major goal of cosmic-ray physics. Observations with ground-based cosmic-ray shower detectors of multi-TeV gamma rays from sources such as Cygnus X-3 have provided a first glimpse of specific sources of cosmic rays. Evidence is fragmentary at present but very exciting. Order-of-magnitude improvements in detection of these signals would allow direct study of particle accelerators at work in nature. On another front, ongoing construction and operation of large underground detectors (originally motivated by the search for proton decay) constitutes a new level of sophistication and collecting power in the study of cosmic-ray muons and neutrinos. At the same time these detectors make possible more-sensitive searches for possible new particles and for neutrinos of extraterrestrial origin. Meanwhile the Fly's Eye detector in Utah is collecting unique data on the highest-energy cosmic rays (above 10^{19} eV).

- New and improved detectors for gamma-ray astronomy in the multi-TeV range
- Continued support of the Fly's Eye and of large underground detectors

II

Gravitation

1

Experimental Tests of General Relativity: Introduction

Perhaps more than in any other area of physics, progress in gravitation physics has been dominated by theoretical work; experimental tests of general relativity have lagged far behind theoretical ideas and predictions. In part this unbalance is due to the extreme difficulty of doing laboratory experiments at interesting levels of accuracy, but it is also true that the elegance and richness of gravitation theory has captured the interest of some of the best theorists of this century. Fortunately for the field, the last two decades have seen dramatic advances in our ability to test gravitation theories. Most of this upsurge in experimental activity was brought about by technological advances in radio and radar astronomy and by the development of precision tracking capabilities for solar-system spacecraft.

The theory of general relativity, devised nearly 70 years ago by Einstein, is still the most successful description of gravitation. Progress in the field has been characterized by the invention of plausible alternatives (such as the scalar-tensor theory) that predict different effects or magnitudes than those predicted by general relativity. Experimental work then decides. Currently, there is no reason to think that general relativity needs modification in the classical domain. As we shall see below, some basic tenets of general relativity have been well tested (parts in 10^{11}), some predicted effects have been measured with good agreement (parts in 10^3), but some major predictions ("magnetic" effects) have not been tested at all.

General relativity makes two distinct statements about the nature of gravitation. First, the metric hypothesis states that gravitation can be described as a Riemannian curvature of space-time, with the laws of physics for all nongravitational interactions having the same form in the local Lorentz frames of curved space-time as in the flat space-time of special relativity. Second, the curvature of space-time is determined, through the Einstein field equation, by the energy, momentum, and stress of all matter and nongravitational fields contained in space-time. Gravitation in this view is an intrinsically nonlinear phenomenon; the field equation alone allows the equation of motion for particles to be deduced from it. This characteristic stands in sharp contrast to Newtonian theory in which the field equation and the equations of motion are separate postulates. Other metric theories of gravitation incorporate the metric hypothesis but differ from general relativity by the manner in which space-time curvature is generated. Experimental tests of general relativity can correspondingly be separated into two categories: tests of the metric hypothesis, such as facets of the principle of equivalence, and tests of the properties of space-time curvature, such as the orbits of light rays and test particles.

The structure of metric theories of gravitation can be clarified by analogy with electromagnetic theory. Gravitation is described by a four-dimensional metric of space-time and electromagnetism by a four-dimensional tensor for the electromagnetic field. However, one often gains insight and computational power by decomposing the four-dimensional quantities into separate spatial and temporal components. In such a decomposition, the electromagnetic field splits into electric and magnetic parts. Similarly, the gravitational field, or metric tensor, separates into three parts: a gravitoelectric field, a gravitomagnetic field, and a part that represents the curvature of space.

In the Newtonian limit of any metric theory of gravitation, the gravitomagnetic field and space curvature vanish; the much stronger gravitoelectric field reduces to the Newtonian gravitational acceleration. In the post-Newtonian regime, a rich variety of new phenomena appear, such as the gravitomagnetic dragging of inertial frames, the gravitoelectric and space-curvature-induced gravitational deflection of light, and the perihelion advance of planetary orbits. To express clearly the consequences of these different post-Newtonian phenomena and the differences between the predictions for each from different metric theories, one can use the parameterized-post-Newtonian (PPN) formalism. With it, all metric theories can be expressed in a common framework in a special coordinate system. In this special coordinate system, the three basic fields—gravitoelectric, gravitomagnetic, and space curvature—are expressed in terms of potentials whose coupling strengths are given by ten dimensionless parameters whose values generally vary from one metric theory to another.

Thus, each theory can be characterized, at this level, by the numerical values of its PPN parameters; and each experiment can be characterized by a predicted result, dependent on one or more of these parameters. Currently the best tested parameters are γ and β; these describe, respectively, the amount of spatial curvature generated by a unit rest mass and the amount of nonlinearity in the superposition of Newtonian gravitational potentials (gravitoelectric fields). There is also one parameter that describes the amount of any preferred-location effect, three that describe the amount and kind of preferred-frame effects, and five (four distinct from those already listed) that describe the amount and nature of violations of global conservation laws for total energy-momentum. An eleventh parameter, $\dot{G}/G$, introduced to describe any fractional time rate of change of the constant of gravitation, depends more on cosmology than on a metric theory of gravitation. For general relativity, γ and β are unity and all other parameters vanish. Although the PPN formalism has its limitations, it has served admirably as a framework to incorporate a large number of theories of gravitation and to stimulate the invention of new experiments.

As we shall see, the best measurements of γ and β have come from experiments using solar-system gravitational fields. The solar system has three special properties in this regard: (a) its gravity is everywhere very weak; the dimensionless ratio of the gravitational potential to the square of the speed of light is 2×10^{-6} on the Sun's surface; (b) the square of the ratio of the speed of each source of significant gravity to that of light is under 10^{-7}; and (c) the ratios of the internal stress energies of all bodies to their respective rest energies are less than 10^{-5}. These three conditions guarantee that Newton's theory of gravitation will provide the same predictions as general relativity to within about 1 part in 10^5 for the structure of the Sun and to within 1 part in 10^6 for experiments confined to the exterior of the Sun. Thus, the goals of most experiments have been to measure deviations from Newtonian theory, i.e., post-Newtonian effects of gravitation whose fractional magnitudes are about 10^{-6} or somewhat less. Of course, higher-order relativistic deviations from Newtonian theory are also predicted to exist in the solar system. These post-post-Newtonian effects are not discernible in present experiments, but they may be reached by the next generation of space experiments. The discovery of neutron stars and perhaps black holes in our galaxy brings hope that experimental gravitation might escape the realm of tiny effects. Multiple systems of these compact objects approach the ideal gravitational laboratory of massive pointlike bodies having negligible nongravitational interactions. One such system—the binary pulsar—has already yielded spectacular results, but the intrinsic advantages of such systems have not yet been fully realized. This remains as a bright hope for the next decade.

2

Experimental Tests of General Relativity: Highlights

This chapter summarizes the current status of tests of general relativity, with emphasis on more recent achievements. For reference while reading this chapter, we list in Table 2.1 the most accurate test results as of mid-1984.

EQUIVALENCE PRINCIPLE, EÖTVÖS TO LUNAR LASER RANGING

In his approach to the theory of gravitation, Einstein did not seek to explain the equivalence of gravitational and inertial mass but instead elevated it to the status of a principle and proposed a generalization stating that, locally, gravitation and acceleration are indistinguishable. The most accurate experimental tests of this principle are of the Eötvös type to determine whether the ratio of inertial to (passive) gravitational mass is the same for all bodies, independent of size or composition. Modern experiments have found no difference in this ratio to a few parts in 10^{11} for several substances. Thus, Eötvös experiments show with high accuracy that nuclear, electromagnetic, and weak interactions contribute equally to gravitational and inertial mass. But does gravitational energy contribute by the same amount?

The gravitational binding energy, important theoretically because it invokes the nonlinear character of gravitation, is too small to measure in laboratory-sized objects. Astronomical bodies must be used, and

TABLE 2.1 Summary of Solar-System Tests of Theories of Gravitation

Measured Effect	Resultant Constraint	Comment
Bound on any non-Newtonian monthly variation in the Earth-Moon distance	$4\beta - \gamma - 3 = 0.001 \pm 0.015^a$	Test of relative contributions of gravitational binding energy to inertial and to (passive) gravitational mass
Comparison between clock in ballistic trajectory and clock on ground	$\dfrac{\text{measured change}}{\text{predicted change}} = 1.0000 \pm 0.0001$	Test of metric hypothesis via gravitational redshift and Doppler shifts
Deflection of radio waves by gravitational field of Sun	$\gamma = 1.01 \pm 0.02$	Test of amount of spatial curvature generated by unit mass
Increase of echo time of radio signals sent from Earth to Mars due to gravitational field of Sun	$\gamma = 1.000 \pm 0.002$	Test of amount of spatial curvature generated by unit mass
Relativistic contribution to advance of perihelion of Mercury's orbit	$(2 + 2\gamma - \beta)/3 = 1.003 \pm 0.005^{a,b}$	Test of combination of amount of spatial curvature generated by unit mass (γ) and nonlinearity in superposition of Newtonian gravitational potentials (β)
Bound on any anomalous acceleration of longitude of planetary orbits	$|\dot{G}/G| \leq 1 \times 10^{-11}\ \text{yr}^{-1}$	Test of constancy of the gravitational constant G

[a] For simplicity in the presentation of results, we have neglected the implied constraints on PPN parameters concerned with violations of global conservation laws and with preferred frame and location effects.
[b] A possible contribution by the solar gravitation quadrupole has been assumed to be negligible.

three or more are required. Our first manned mission to another astronomical body enabled an accurate test to be performed with the Earth-Moon-Sun system. Emplacement on the lunar surface of optical corner reflectors by the Apollo astronauts has allowed us to distinguish whether the Moon and the Earth fall toward the Sun with equal accelerations. Any anomalous difference in these two accelerations would manifest itself in a corresponding monthly variation in the Earth-Moon distance, now determined from laser measurements to within 10 cm. The measurements set stringent limits on any anomalous behavior and establish that at least 98.5 percent of the gravitational binding energy of the Moon contributes to both its gravitational mass and its inertial mass. To this accuracy, therefore, it has been verified that all ordinary mass-energy, including that due to gravitational self-energy, gravitates in the same manner. This result constrains a combination of PPN parameters; for the special case of fully conservative metric theories without preferred frame or location effects, it implies that the linear combination $4\beta - \gamma - 3$ vanishes to within ± 0.015. Some metric theories predict a violation of the principle of equivalence for massive bodies because, in these theories, only part of the mass due to gravitational self-energy gravitates, although the principle is obeyed for the contributions to mass from all other forms of energy. The class of such theories has thus been sharply curtailed by this result from the lunar laser-ranging experiment.

Space techniques may provide an opportunity for improving the classical Eötvös experiment. An apparatus is being developed where two masses (of different composition) in the form of concentric cylinders are free to move along their common axis on magnetic bearings. In orbit around the Earth the difference in their free-fall accelerations would be measured. The geometry minimizes the effect of gravity gradients, which are large for a torsion balance experiment. It is anticipated that ground-based tests with this apparatus should reach an accuracy of 10^{-12}, and in space the experiment may reach an accuracy of 10^{-15}, depending on the levels of mechanical and gravity gradient disturbances.

GRAVITATIONAL REDSHIFT, MÖSSBAUER TO ROCKETBORNE MASER

One of the most celebrated predictions of general relativity concerns the effect of gravitational potential on the rates of clocks and on the frequency of an electromagnetic signal. A given clock appears to run more slowly than an identical clock located in a region of lower

18 GRAVITATION

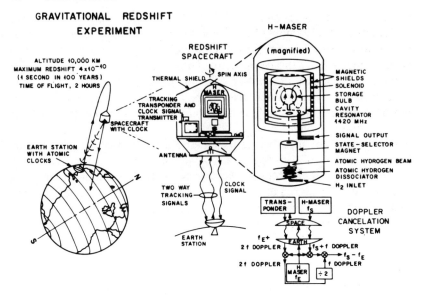

FIGURE 2.1 A suborbital clock has measured the gravitational redshift effect; the result agrees with theory to within the experimental accuracy of 1 part in 10^4. Keys to the success of the experiment were the special hydrogen maser and the two-way communications link that allowed subtraction of a huge Doppler effect.

gravitational potential. The most precise laboratory verification of the gravitational redshift effect was obtained a decade and a half ago using the Mössbauer effect to obtain extremely narrow spectral lines. By velocity compensation of the change in frequency of the gamma rays over a vertical distance of 25 m, it was possible to verify the prediction of general relativity to about 1 percent.

By far the most accurate experiment to test the effect of gravitation on the rate of a clock was performed by the placement of a hydrogen-maser frequency standard on a rocket that traveled on an orbital arc with a 10,000-km maximum altitude. In this experiment, diagramed in Figure 2.1, a sophisticated radio communication link was employed to circumvent ionospheric propagation effects and to cancel the large Doppler shift. Thus, the rate of the hydrogen-maser clock in orbit is accurately compared with similar masers on the ground. The measured redshift agreed with the prediction to within the experimental uncertainty of about 1 part in 10^4—the most accurate relativity experiment yet performed with space techniques.

LIGHT DEFLECTION, ECLIPSES TO RADIO INTERFEROMETRY

Electromagnetic radiation is predicted by general relativity to be deflected by massive bodies, in part from the action of the gravitoelectric component of the gravitational field (a direct consequence of the principle of equivalence) and in equal part as a consequence of space curvature.

The deflection of light by the Sun was dramatically verified by an eclipse expedition team in 1919, catapulting Einstein to world fame. But Earth-based observations of total eclipses have not achieved the level of reliability needed for accurate verification of the predicted deflection. In the late 1960s optical eclipse observations were largely supplanted by radio-interferometric techniques. Simultaneous measurements at two radio-frequency bands enable the refractive effects of the solar corona to be reduced to a benign level. As a result, the uncertainty of the verification of the predicted 1.75-arcsec deflection for rays grazing the solar limb was decreased by a factor of 10, now implying that γ is unity to within about 2 percent.

SIGNAL RETARDATION, NEWEST AND MOST ACCURATE TEST

General relativity also predicts that the transit times of electromagnetic signals traveling between two points will be increased if a massive body is placed near the path of these signals. Thus, a measurement of the round-trip time of signals propagating between two points will be greater the nearer a massive body lies to the path of propagation, owing in part to the principle of equivalence and in equal part to space curvature, as for light deflection. The development of radar and of space techniques made this test possible, and as a latecomer it is sometimes called the "Fourth Test," the classical three being Mercury's perihelion precession, light deflection, and the gravitational redshift. Signal retardation measurements currently provide our best test of the important space-curvature effects in general relativity. The increase of the round-trip times for light or radio signals propagating between planets, owing to the direct effect of solar gravitation, is predicted by general relativity to reach a maximum of about 250 μs for ray paths that graze the limb of the Sun. This prediction was verified first through measurement of echo times of radar signals bounced from the surfaces of the inner planets.

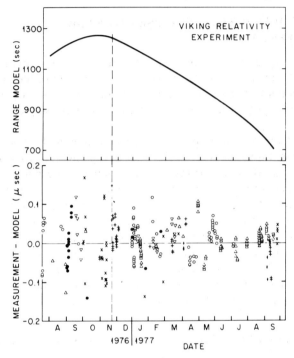

FIGURE 2.2 Ranging to the Viking Landers. Shown here are the residuals after fitting measured round-trip times to the range model. Measurement uncertainties are omitted to avoid cluttering the figure. Mars was on the other side of the Sun on November 25, 1976. VL1 and VL2 denote Viking Landers 1 and 2, and 14, 43, and 61 and 63 denote, respectively, Deep Space Network tracking stations in Goldstone, California; Canberra, Australia; and Madrid, Spain (26- and 64-m antennas).

More recently a 50-fold improvement in this test was realized by using the Viking Lander spacecraft on the surface of Mars. The round-trip travel times of radio signals were measured with uncertainty as small as 10 ns, about 10^{-11} of the total travel time. The measurements were then fit to an elaborate range model including many solar-system parameters, the relativistic delay, and positions for the spacecraft and tracking stations. The residuals of the measurements from the model are shown in Figure 2.2. The final uncertainty in measuring the relativistic delay arises from possible systematic errors and parameter correlations in the model fitting. The Viking experiment reduced the uncertainty in the measurement of the relativistic delay from 5 percent (obtained with radar) to 0.1 percent. The measured delay agrees with the prediction of general relativity [which is proportional to $(1 + \gamma)/2$], showing that $\gamma = 1 \pm 0.002$—an order of magnitude higher accuracy than yet achieved for the light deflection test.

PERIHELION ADVANCE, EINSTEIN'S ONLY HANDLE

The anomalous advance of the perihelion of the orbit of the planet Mercury, noted in the mid-nineteenth century, provided the first hint that Newtonian theory was not adequate as a description of the dynamics of the solar system. This advance, subsequently determined to be 43 arcsec per century, was an elegant confirmation of Einstein's theory. Because this effect increases secularly, the improvement from use of modern radar observations of Mercury over the results obtained from several hundred years of optical observations has not been so dramatic. At present, radar observations of Mercury yield an uncertainty of 0.5 percent in the determination of the anomalous perihelion advance, a twofold improvement over the results from optical observations. The relativistic contribution to the perihelion advance depends not only on space curvature but also on the nonlinearity of the superposition law for the gravitational potential and on preferred-frame and location effects. If one assumes that the contributions of the solar quadrupole moment and of possible preferred-frame and location effects are negligible, the measurements demonstrate that for fully conservative theories the combination $(2 + 2\gamma - \beta)/3$ of PPN parameters is unity to within 0.5 percent. Relativistic perihelion advances have also been detected for Mars and for the asteroid Icarus. The results agree, to within the 20 percent experimental uncertainties, with the values predicted by general relativity.

CHANGING GRAVITATIONAL CONSTANT, SOLAR-SYSTEM TIME VERSUS ATOMIC TIME

A deep question of physics concerns possible variations with time of certain constants of nature. General relativity assumes that the constant of gravitation G is a universal constant, independent of both spatial location and time. The possibility that this constant varies with time is based in part on the so-called large numbers hypothesis. This hypothesis stems from the fact that the ratio of the electrostatic to the gravitational force between an electron and a proton, about 10^{39}, is approximately equal to the age of the universe expressed in atomic units. Is this near equality a mere coincidence confined to the present epoch? If one assumes instead that it is of fundamental significance, independent of epoch, then some physical constant must vary with time. It has been proposed that the gravitational interaction, as measured against the electromagnetic, may be weakening with time. Any such effect should be detectable by comparing time kept by an atomic clock with the time kept by a gravitational clock. In practice, precise ranges to solar system bodies are measured as a function of time, as kept by atomic clocks. The ranges are fitted to an elaborate solar-system model that includes general relativity and a possible effect due to a changing value of G. Recent results from ranges to Mars (using the Viking Lander and Mariner 9), radar ranges to Mercury and Venus, lunar laser ranges, and optical positions of the Sun and planets have set a limit at about $|\dot{G}/G| < 10^{-11}$/year. Accuracy is limited more by incompleteness of the solar-system model than by experimental errors; currently the limit is imposed by uncertainty in the gravitational perturbation of Mars by the asteroids.

LABORATORY TESTING OF GRAVITATION, SEARCHING FOR THE UNEXPECTED

We must not allow these impressive advances afforded by space techniques to overshadow completely the important contributions of laboratory gravitation experiments. Many of these experiments achieve great accuracy by using null techniques, as in the celebrated Eötvös experiments. The methodology is to propose plausible anomalies to the standard theory or its assumptions. Null experiments are then devised such that the proposed anomaly leads to a nonzero result. Because of the characteristic high precision of null experiments, the results often yield deeper and broader insights than originally intended.

Many basic aspects of Newtonian gravitation are taken for granted in spite of a lack of experimental verification. Recently, the validity of the R^{-2} dependence of gravitation for laboratory distance scales has been questioned and tested. (From 10^4 km to planetary distance scales the exponent is known to be -2 with an accuracy of a few parts in 10^8.) Torsion balance experiments give an exponent of $-2(1 \pm 0.1$ percent) on distance scales from a few centimeters to a meter, while surface and satellite measurements of the Earth's gravitational field give $-2(1 \pm 1$ percent) on a 1-km scale. Although the results are not surprising, they do put gravitation on a better footing. An unexpected bonus of the short-range experiments is that the results place constraints on properties of possible new particles (e.g., axions) that might lead to short-ranged exchange forces in ordinary matter.

When we write down Newton's second law for a planet orbiting the Sun we generally do not notice that the three masses in that equation are playing distinctively different roles. The active (attractor) mass, passive (attracted) mass, and inertial mass are assumed to have the same ratios, independent of composition. This is a fundamental assumption of general relativity and is well tested for passive and inertial masses where large solar-system bodies can be used as the active third mass. Unfortunately, experiments to compare active mass with inertial or passive masses necessarily use laboratory-scale masses. One technique uses a Cavendish-type experiment except that the large movable (active) mass floats beneath a fluid of exactly the same (passive) density. As the mass moves back and forth, the torque on a torsion pendulum is proportional to the difference in the ratios of active to passive masses for the solid mass versus the displaced fluid material. A composition dependence in (active mass)/(passive mass) would result in a nonzero torque. The ratio has been found to be the same for fluorine and bromine to an accuracy of a part in 10^4.

Laboratory experiments to look for effects of local anisotropy of space have achieved high precision. These so-called Hughes-Drever experiments search for tiny frequency shifts in atomic and nuclear resonance lines that might be correlated with the orientation in space of a polarized nucleus, rotated once a day by the Earth. Exceedingly small shifts (compared to nuclear binding energy) are detectable, leading to one of the most accurate null results in physics: inertial mass is locally isotropic to better than 10^{-20}. Though more than two decades old, we mention these important results because new techniques in atomic physics have brought renewed interest in the experiments. Only a few experimenters choose to do laboratory gravitation. The work is characterized by clever techniques, compulsion with systematic errors, long integration times, and great experimental ingenuity.

3

Experimental Tests of General Relativity: Opportunities

TESTS FOR "MAGNETIC" GRAVITATIONAL EFFECTS

At present there is no experimental evidence arguing for or against the existence of the gravitomagnetic effects predicted by general relativity. This fundamental part of the theory remains untested. The reason is simple; predicted effects, such as the dragging of inertial frames by rotating massive bodies, are exceedingly small near solar-system bodies (though they can be enormous and astrophysically crucial near a rotating black hole). The precision solar-system experiments described above probe space-curvature effects and the gravitoelectric field, but the predicted effects due to rotation of the Sun and Earth are too small to be detectable in experiments performed to date.

Relativity Gyroscope Experiment

An experiment has been devised to search specifically for the frame dragging effect. NASA's Relativity Gyroscope Experiment (Gravity Probe B, see Figure 3.1) will use test gyroscopes in orbit to look for frame dragging by the rotating Earth. A test gyroscope defines the orientation of the local inertial frame, and the experiment looks for a precession of this frame with respect to the fixed stars. The main difficulty is to reduce external torques on the gyroscope to an excep-

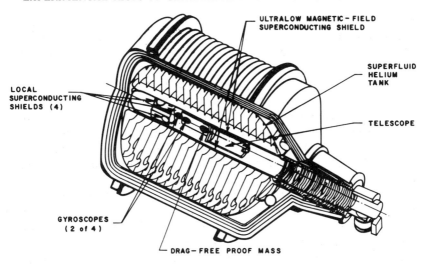

FIGURE 3.1 The Relativity Gyroscope Experiment is our best hope of testing the unexplored magnetic-like effects in general relativity. In polar orbit, the telescope will be accurately pointed to a reference star, and the precession rates of the precision gyroscopes will be monitored to an accuracy of a few milliarcseconds/year.

tionally low level; otherwise they would induce mechanical precession, which masks the tiny precession due to frame dragging.

The most interesting precessional effect predicted by general relativity goes by several names: motional, frame-dragging, Lense-Thirring, and gravitomagnetic among others. As a consequence of coupling between the gyroscope spin and the rotating Earth, the effect is analogous to the spin-spin coupling that gives rise to atomic hyperfine spectra. For an orbital altitude of about 600 km above the Earth, the maximum frame-dragging precession is 0.044 arcsec/year; thus, the design goal for the experiment is a precision of 0.001 arcsec/year. General relativity also predicts a geodetic precession of 6.9 arcsec/year, which is split between two physical effects. There is a spin-orbit precession where the gyroscope spin couples to the gravitomagnetic field induced in the gyroscope's rest frame by its motion through the Earth's gravitoelectric field. This amounts to 2.3 arcsec/year. The remaining 4.6 arcsec/year arises from the gyroscope's motion through the curved space near the Earth. Neither the frame dragging nor the geodetic precession has been directly observed in any past experiment.

Gravity Probe B is planned around four identical gyroscopes and a reference telescope, all fabricated from fused quartz and kept at a temperature of 1.6 K. Each gyroscope will consist of a quartz sphere almost 4 cm in diameter, coated with a superconducting niobium film and suspended electrostatically. The initial spin rate of nearly 200 revolutions per second is expected to decay by less than 0.1 percent during the course of a year because of the very low (10^{-10} Torr) pressure maintained within the vessel. To reduce external torques from the suspension and from gravity gradients, each gyroscope rotor has to be round to better than 1 part in 10^6 and homogeneous to within a few parts in 10^7. The orientation sensor uses the sphere's London moment and low-noise superconducting quantum interference device (SQUID) magnetometers to read the spin-vector alignment with the necessary precision and without exerting significant sensing torques on the gyroscope. Superconducting lead bags are used to reduce residual dc magnetic fields to below 10^{-7} gauss. The spacecraft uses a drag-free proof mass to reduce nongravitational accelerations on the gyroscopes to about 10^{-7} cm/s^2. To modulate the precession signal, and to average out some unwanted torques, the spacecraft is slowly rolled. The telescope views a bright reference star (probably Rigel) along the roll axis; the proper motion of this star will be determined from separate observations.

Clearly, this is an exceedingly difficult experiment, many times more sophisticated than any yet attempted in space. Some of the critical technology is new and therefore of higher risk than is usually considered prudent for space experiments. Yet, the experiment has withstood intensive technical reviews, which found that a successful experiment is possible, if done with care. Scientific reviews have always been enthusiastic because the science is compelling, and the experiment is unique.

NASA's current plan is to develop the experiment in two stages. Stage 1 will consist of building the flight Dewar and instrument, including all four gyroscopes, and performing an engineering test in the relatively low-g environment of the Shuttle spacecraft. In stage 2 the refurbished instrument will be flown in a free-flying spacecraft to obtain the ultralow-g environment required for the experiment. This approach is designed to minimize the risk associated with the experiment's advanced technology.

Black-Hole Jets

It is possible that astronomers may now be seeing a very dramatic gravitomagnetic effect. A few quasars and strong radio galaxies exhibit long jets of gas and associated magnetic field emanating from their nuclei. Some jets are surprisingly straight, requiring good alignment of the source for $\sim 10^7$ years. Others show corkscrew patterns, suggesting precession with periods of $\geq 10^4$ years; and others are more complicated. A plausible current theory is that the sources of these jets are rotating supermassive black holes, $M \gtrsim 10^7$ solar masses, in the nuclei of some galaxies: the gyroscopic action comes from the hole's rotation-induced gravitomagnetic field, and the corkscrew jets may result from geodetic precession of the hole's spin as it orbits around another massive body. How might a black hole generate a collimated, energetic jet? One exotic but physically plausible mechanism relies on the dipole-shaped gravitomagnetic field of a rotating black hole. That field, derived from the interaction of the hole's horizon and the magnetic field deposited on the hole by a surrounding accretion disk, drives charged particles away from the hole's poles in ultrarelativistic beams. The energy ultimately comes from the black hole's rotation. Figure 9.2 in Chapter 9 depicts this model. Unfortunately, the complexity of such a system, and the poor prospects for getting detailed data, make it unlikely that observations of jets will ever constitute a quantitative test of gravitomagnetism in general relativity.

RANGING TO THE MOON AND INNER PLANETS

For the coming decade, range measurements to the Moon and inner planets will continue to provide important tests of general relativity. Ranges are currently being measured with the exquisite accuracy of 1 part in 10^{11} in some cases, and as we see in Table 2.1, the scientific payoff has been outstanding. But we can do even better by pushing the measurements to the technically feasible limits and by scheduling observations for best scientific advantage.

Before discussing specific possibilities we should point out two important characteristics of solar-system range measurements.

1. The whole is much greater than the sum of its parts. At the levels probed, the solar system is a complex network of gravitational interactions, modeled by an elaborate ephemeris. Each experiment couples to this network with its own unique matrix, and often the interrelations of different experiments are important but by no means apparent. Furthermore, many effects (such as Mercury's perihelion precession) are cumulative with time, so measurements made over the long term are especially sensitive. For these reasons analysis of the total available data set can enhance the reliability and accuracy of any single test

of a theory of gravitation, and data obtained during one space mission might be of only moderate value in themselves but, when combined with data taken in another, might be of great interest.

2. Measurements of the dynamics of the solar system, made with modern instrumentation, will be an extremely valuable legacy to leave to future generations of scientists, who will combine their data with those obtained in the present era and reap more sensitive tests of the fundamental theories of gravitation. The history of gravitation physics provides a shining example of the importance of such legacies. The observational work of Tycho Brahe, its use by Kepler, and the work of many generations of observational astronomers enabled Leverrier in the mid-nineteenth century to detect the anomalous advance in the perihelion of Mercury's orbit, later to become general relativity's first successful test.

Radar Ranging

Radar ranges to Mercury currently provide our best measurements of the perihelion precession predicted by general relativity. The uncertainty in the determination of the total perihelion advance decreases as the $-3/2$ power of the time interval spanned by the data, so a long-term program is important. Given the current infrequency of planetary spacecraft missions (see below), it is particularly important to maintain and improve our radar capability.

Sustained high-accuracy measurements of the echo delay of radar signals between the Earth and the inner planets are being accomplished at present with the NASA-supported radar facilities at the Arecibo Observatory and at the Goldstone Tracking Station. The main limitation on the utility of such data for tests of relativistic gravitational effects has not been measurement accuracy but rather measurement sparsity and the unknown topography of the target planets. Increasing the frequency of measurements and exploiting techniques to map planet topography can substantially improve the radar-ranging contributions to this field.

Ranging to Planetary Landers and Orbiters

Range measurements from the Earth to the Viking Landers on Mars have been particularly valuable in testing gravitation theories. Ranges were measured with uncertainties as low as 3 to 5 m near opposition, the highest fractional accuracy achieved so far in solar-system measurements. However, the Viking Landers are no longer in operation, and there are no specific plans at present for future U.S. landers on any of the planets. The most likely location for a future lander would be Mars, and the possibility of ranging to such a lander with an overall measurement accuracy approaching 1 cm should be pursued actively. The striking scientific success of the Viking Lander tracking measurements provides a strong justification for obtaining range measurements to future landers and for increasing the accuracy as much as possible.

In view of the infrequent opportunities that are likely to arise for ranging to planetary landers, it is important to utilize improved techniques for ranging to orbiters to obtain high-accuracy planetary distance measurements. This is particularly desirable for Mercury for several reasons. One is the greatly increased accuracy with which the precession of Mercury's perihelion could be obtained. Several years of high-accuracy radio-tracking data would give an independent measurement of the solar quadrupole moment, allowing separation of the relativistic precession and the precession due to the solar quadrupole. A Mercury orbiter also offers good prospects for lowering the present upper limit on $|G/G|$ of 10^{-11} per year by several orders of magnitude. This is partly because of improvements in measurement accuracy and partly because asteroid perturbations are smaller for Mercury and the Earth than for Mars.

The main limitation on obtaining interplanetary distances by ranging to planetary orbiters comes from uncertainty in the spacecraft orbit with respect to the planet's center of mass. This uncertainty, in turn, stems in large part from a lack of knowledge of the planet's gravitational field. For this reason, the use of a relativity subsatellite in a fairly high-altitude orbit, with a small eccentricity, is most favorable. Tracking of such a satellite simultaneously at two radio-frequency bands, say the X band and the K band, should allow removal of virtually all uncertainties in distance measurements, and in radial velocity measurements, due to interplanetary plasma.

The first major opportunity to utilize a planetary orbiter will be through the Mars Observer Mission. Such an opportunity, of interest in its own right, would also enable the refinement of the techniques proposed for use with Mercury orbiters. Determination of the gravity field could be accomplished via use of a dual-frequency tracking system similar to the system incorporated in the Galileo spacecraft. Inclusion of an accurate ranging system would allow the Mars Observer itself to be utilized to improve on the spectacular results for testing general relativity obtained from the Viking Landers on Mars.

30 GRAVITATION

FIGURE 3.2 The array of corner reflectors placed on the Moon by Apollo 14 astronauts (note footprints). The bubble level and gnomon, used for pointing the array toward the Earth, can be seen. Laser range measurements are routinely made to three widely separated arrays. No degradation of their optical reflectivity has been observed.

Lunar Laser Ranging

Laser range measurements to optical corner reflectors on the Moon (see Figure 3.2) have been made for over a decade with an uncertainty of about 10 cm from 20 minutes of observation. Recently, additional sites in Hawaii and Australia have joined the McDonald Observatory in Texas and the Grasse Observatory in France in making regular range measurements. The accuracy from the new stations and, after improvements, from the older stations is expected to be a few cm. We expect that the equivalence principle test (does gravitational binding energy have inertial mass?) will be improved tenfold over the current accuracy. [This test already provides our best accuracy for measuring the parameterized-post-Newtonian (PPN) parameter β.] Geodetic precession of the lunar orbit (with the Earth-Moon system playing the role of a gyroscope in orbit around the Sun) might also be determined to about 10 percent of the predicted effect. This accuracy, however, is far lower than is expected from the Relativity Gyroscope Experiment (GP-B) discussed earlier. Additionally, these laser-ranging data should allow an accurate measurement of a possible change in the gravitational constant, because the Earth-Moon tidal acceleration is being measured independently by LAGEOS ranging experiments. Laser observations are also useful for a variety of applications in geophysics and selenophysics such as the determination of Earth rotation and nutation, the lunar mass distribution, and the excitation of free libration of the Moon.

Finally, we emphasize the important interrelationship between planetary and lunar-ranging measurements. The combination of the equivalence principle test from lunar ranging with information on the planetary mean motions, perihelion precession, and time delay from planetary ranging strengthens our present ability to set a limit on the solar quadrupole moment and to determine other important solar-system parameters such as GM_{Sun}. Also, the planetary observations aid the analysis of lunar-ranging data. Thus, the contribution of any given set of measurements must be judged not in isolation but in regard to its effect on deductions from the ensemble of measurements, past as well as future. It is for this reason that each feasible opportunity for ranging to the Moon and planets should be seized.

MEASUREMENT OF SECOND-ORDER SOLAR-SYSTEM EFFECTS

All past measurements of nonlinearity in the superposition of gravitational potentials (PPN parameter β) have involved the dynamical motions of test bodies such as Mercury, whose perihelion precession rate agrees with that predicted by general relativity. A high-precision clock experiment would probe β in a different physical context and would check whether, at a nonlinear level, gravitation can be represented by a metric theory. It has been proposed to put a hydrogen-maser clock aboard a solar probe spacecraft, called STARPROBE, which would travel in an eccentric, near-Sun orbit. Such a mission would provide a superb gravitational redshift experiment as well as making the first clock measurement of β. The change in gravitational potential is 10^3 larger than that experienced by the rocketborne hydrogen maser, which holds the record for gravitational redshift tests, 1 part in 10^4. To measure β with 10 percent accuracy requires a clock stability of 1 part in 10^{15}, or better, for averaging times of 10^2 to 10^6 seconds. Comparable performance has been achieved in the laboratory for averaging times up to 10^4 seconds. Development of a spaceborne experiment requires careful environmental control to accommodate extreme solar heating. During its several-year cruise to the Sun, a solar probe with an ultrastable oscillator on board would offer an unprecedented opportunity to search for long-period gravitational waves, as discussed later in the section on Pulsar Timing and Millisecond Pulsars.

Another space project has been proposed to test general relativity to unprecedented levels of accuracy—three orders of magnitude more sensitive than present solar-system tests. The idea is to measure solar deflection of starlight with sufficient accuracy to detect the second-order contribution of the gravitational potential, a deflection of 10.9 μarcsec at the solar limb. The instrument envisioned is an articulated pair of stellar interferometers with their viewing axes approximately 90° apart. The instrument (called POINTS, an acronym for Precision Optical INTerferometry in Space) would have two pairs of mirrors of 1-m diameter and an interferometer separation of 10 m; statistical accuracy after 5 min of integration on 10th-magnitude stars is under 1 μarcsec. The challenging problem of achieving absolute accuracy appears to be solvable by means of internal laser-beam metrology.

Figure 3.3 shows a smaller version of the interferometer that could fit fully assembled with a supporting spacecraft into the Shuttle bay. This instrument would have 25-cm mirrors separated by 2 m. For a pair of 10th-magnitude stars, it would measure the separation with a statistical uncertainty of 5 μarcsec after a 15-min observation. Although possibly falling short of the accuracy needed for a second-order test, this interferometer would allow at least a 2-order-of-magnitude improvement to be made in the accuracy of the solar light-deflection experiment. Such an experiment could be conducted from the bay of the Shuttle and provide an estimate of the PPN parameter γ ten times better than did the Viking Lander mission using the time-delay test.

POINTS also has obvious applications in precision astrometry. Parallax and proper motion studies could be extended to all visible parts of the galaxy, contributing to our understanding of the cosmic distance scale and galactic dynamics. Statistical studies of the abundance of planetary systems should be possible.

Although the decision to develop a space-based astrometric instrument must be based on the predictable scientific results of such a

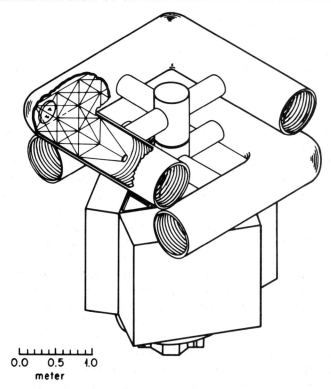

0.0 0.5 1.0
meter

FIGURE 3.3 An artist's rendition of a small optical interferometric satellite to be used for precision astrometry. It consists of two U-shaped interferometers joined by a bearing that permits the angle between the principal axes of the interferometers to vary by a few degrees around its nominal value of 90°. Each telescope has a 25-cm-diameter mirror and is separated by 2 m from its companion. NASA's Multimission Modular Spacecraft is shown mounted under the instrument.

mission, the most important results may be the serendipitous discoveries that seem to follow when a new instrument provides a large set of observations that are orders of magnitude more accurate than previously available. Further studies of such an optical interferometer are required now to prepare for an eventual space mission.

GRAVITATIONAL QUADRUPOLE MOMENT OF THE SUN

A solar quadrupole moment causes the perihelion of Mercury's orbit to precess, and uncertainty in the magnitude of this effect has been a long-standing problem for the relativistic interpretation of the measured precession. A large quadrupole moment could be caused by rapid rotation of the solar interior; however, a uniformly rotating Sun has a small quadrupole and negligible effect on Mercury's perihelion at the present level of measurement accuracy. Despite considerable effort, neither solar-system tracking experiments nor ground-based optical oblateness experiments have convincingly measured or ruled out a solar quadrupole effect.

The recent discovery of high-Q solar oscillations with periods near 5 min has introduced a new method of indirectly determining the solar gravitational quadrupole moment. These modes of oscillation have radial extent going deep into the Sun, so rotational splitting of the mode-frequency structure is being used to probe the rotation rate of the solar interior. Knowing the radial dependence of rotation, the solar model can be used to calculate the gravitational quadrupole moment. Early results of this method indicate a value close to that for uniform rotation of the Sun, with small quoted uncertainty. Work is under way on better observations, which will include spatial resolution, and on more detailed models relating mode structure and the solar interior.

We have already noted (see section on Ranging to Planetary Landers and Orbiters) that accurate radio tracking of a satellite orbiting Mercury would give a much more accurate measurement of the perihelion precession, as well as a direct measurement of the solar quadrupole moment. The direct measurement would not only support a more accurate perihelion measurement, but would also be an important check on our understanding of the solar interior and of solar oscillations.

SYSTEMS OF COMPACT STARS

Multiple systems of neutron stars and/or black holes present new opportunities for research in gravitational physics. The ideal of studying a system of pointlike objects of large mass with negligible nongravitational interactions was just a dream before the discovery and close study of the binary pulsar system PSR 1913+16. This is a 16-Hz pulsar in an 8-h orbit around an unseen companion. Pulse timing data of high precision allow unprecedented scrutiny of many orbit parameters, including four relativistic effects—periastron precession rate, gravitational redshift, transverse Doppler shift, and orbital decay due to gravitational radiation. As our first evidence for the existence of gravitational radiation, we will highlight this system in Chapter 5 (in the section on Sources of Gravitational Waves—Recent Developments). Here we are concerned with the potential of such systems as astrophysical laboratories for testing other predictions of general relativity.

The binary pulsar is an almost ideal gravitational laboratory. Large orbital eccentricity (0.617127 ± 0.000003), small orbital size ($a\sin i$ = 2.34185 ± 0.00012 light seconds), and large masses (each mass = 1.41 ± 0.03 M_{Sun}) lead to relatively large gravitational effects. The periastron precession rate is found to be 4.2263 ± 0.0003 degrees/year compared with 43 arcsec/century for Mercury. Why is this measurement of periastron precession to 2 parts in 10^4 not listed in Table 2.1 instead of Mercury's precession (5 parts in 10^3)? The reason is that pulsar timing data do not independently give the masses of the pulsar and companion, and these are needed to calculate the size of the relativistic precession. Instead the measured precession rate is used to find the masses (including the first high-precision mass measurement for any neutron star), assuming that general relativity is correct. The model and data are all self-consistent and strongly suggest that we are observing a clean gravitational system with relativistic periastron precession and gravitational radiation. Nevertheless, the possibility remains that the companion might be a helium star or white dwarf. Calculations indicate that these objects could conceivably have a mass quadrupole moment large enough to cause the observed periastron precession and/or tidal dissipation sufficient to cause the observed orbit decay. Thus, the agreement of the measurements with the predictions of general relativity could be fortuitous.

The binary pulsar is a breakthrough in gravitation physics. By exhibiting large gravitational effects, such systems offer exciting opportunities for testing general relativity. Suppose, for example, that the companion in PSR 1913+16 had turned out to be a pulsar; neutron stars are sufficiently pointlike that no ambiguity would remain in interpreting the measurements. One can imagine other systems similar to this one where pure gravitational interaction could be shown with certainty to dominate the dynamics. Systematic searches for compact star systems and detailed measurements of their properties should be vigorously pursued whenever possible.

4

Search for Gravitational Waves: Introduction

General relativity theory can be tested on Earth and in the solar system only through its weak-field, slow-motion effects. When gravitational fields become strong, and when matter velocities approach the speed of light, new phenomena occur. A black hole, formed by gravitational collapse of a stellar core, is one example. Another is a wave in the space-time metric, traveling at the speed of light—the gravitational wave. Gravitational waves interact only weakly with matter and thus are hard to detect. The detection of gravitational waves is the most important unsolved problem in experimental gravitation today. Their detection would provide an important test, in a new regime, of Einstein's general theory of relativity and might also open a new astronomical window and give new kinds of information about the sources of gravitational waves. Intriguing possible sources are collapsing stellar cores, colliding neutron stars or black holes, decaying binary systems, rotating or vibrating neutron stars, and new sources of unknown nature.

Development of several kinds of gravitational-wave detector has continued for two decades, with great advances in technology, but with no discovery as yet. Current trends in technology of the detectors, together with the best theoretical guesses of strength and event rate of astronomical sources, lead one to anticipate that gravitational waves may be detected within the next decade or two.

Meanwhile, the discovery and long-term observation of a radio pulsar in a binary stellar system has provided impressive evidence that gravitational waves do exist. The orbit of this system is decaying at just the rate expected owing to gravitational-wave damping.

THEORY

In any theory of long-range forces that is consistent with special relativity, the force must act at the speed of light rather than instantaneously. Consequently there is a strong expectation that, along with the static long-range gravitational force, there must exist in nature some kind of gravitational-field excitation that travels at the speed of light and that can remove energy from an isolated system—gravitational radiation or gravitational waves.

Einstein himself showed the existence of gravitational waves in the general theory of relativity, soon after the theory was complete. However, he used the linear approximation to general relativity in deriving this result, and the fact that general relativity is intrinsically a nonlinear field theory led many to doubt the existence of waves. For about 40 years confusion reigned on the issue of whether gravitational waves were or were not a prediction of general relativity, and the theoretical issue was settled only in the early 1960s. The theoretical properties of gravitational waves, presuming the correctness of general relativity theory, are now thought to be well understood.

Alternative theories of gravity usually also predict gravitational waves, although with significant differences from the predictions of general relativity. In some such theories (either those with prior geometry or with more than one metric tensor) the speed of gravitational waves may differ from the speed of light and from the speed of all other massless particles. The difference typically depends on the ratio of the gravitational potential to c^2 and amounts to about 1 part in 10^6 for gravitational waves traveling in the gravitational field of our galaxy. But this already amounts to a difference of arrival time of several days between the gravitational-wave pulse and the neutrino or photon pulse from, say, a supernova in our galaxy, and a greater difference for extragalactic sources. Alternative theories also generally predict different polarization properties for gravitational waves. This is because general relativity contains only a spin 2 (tensor) field, while other theories typically also incorporate scalar fields of spin 0 or vector fields of spin 1. Therefore general relativity theory predicts only quadrupole deformations of a gravitational-wave antenna, while other theories predict monopole or dipole deformations as well.

SOURCES

Because of the weakness of the gravitational interaction, it seems impossible to create on Earth a source of gravitational waves strong enough to be sensed by any conceivable detector; this means that it is impossible to carry out the gravitational analog of Hertz's experiment, and we must depend on cosmic sources to excite detectors.

Astrophysical phenomena involving the coherent motions of large, compact masses at relativistic speeds are the sources most likely to emit measureable gravitational radiation. It is, however, just these extreme phenomena that, if they can be observed, will allow us to test relativistic gravitation in the strong-field, high-velocity regime. A view held by many is that this is the most important reason to engage in the search for gravitational radiation. The signatures of gravitational waves may well be the most definitive means to establish the existence of black holes and to study the interactions of compact objects of all kinds with their surroundings. Thus, the detection of gravitational radiation has become an important problem in relativistic astrophysics.

Estimates of the gravitational-wave spectrum incident on the Earth suffer from our limited knowledge about massive compact objects in the universe. If the precedent set by the development of radio, infrared, and x-ray astronomy serves as a guide, chances are excellent that the first sources of gravitational waves to be detected will not have been included in the present inventory of hypothesized sources. Several classes of known astrophysical objects have been proposed as emitters of gravitational radiation. A few of these are described below, and estimates of their strength at the Earth are shown in Figures 6.2-6.4 in Chapter 6.

The collapse of stellar cores in Type II supernovae may produce millisecond bursts of gravitational radiation provided there is sufficient departure from spherical symmetry in the collapse. A supernova at the center of our galaxy, if it released 1 part in a thousand of its total mass into gravitational waves, would produce strains* of the order of 10^{-18} at the Earth. Such a strain measurement is just barely within the capabilities of currently operating detectors. The supernova rate in our

---

* A passing gravitational wave causes two freely falling masses to undergo relative acceleration and a displacement proportional to their separation. Similarly, a strain is induced in a solid body. Thus, the strength of a gravitational wave is customarily measured by the displacement per unit separation, or strain h. This quality is also equal to the perturbation in the space-time metric accompanying the wave.

galaxy is however only about 1 per 10 years. To gain event rates of a few per year one must reach out to the Virgo cluster of galaxies with strain sensitivities of 10^{-21}. Detectors having such a sensitivity would be able to detect supernovae in our own galaxy in which only 10^{-9} of the mass is converted to gravitational radiation.

Neutron stars in binary systems gradually spiral together owing to the emission of gravitational radiation. The binary pulsar PSR 1913+16 is an example of such a system. In the final hours of its existence the binary system will emit a strong chirp of gravitational radiation sweeping from 10 Hz to 1 kHz, terminated by the tidal disruption of one or both of the stars themselves. The event in PSR 1913+16 would produce strain amplitudes of 10^{-18} at the Earth, but we will have to wait about 10^8 years for this to occur. By inferring a death rate for such binary systems from pulsar observations, one can anticipate that detectors having a strain sensitivity of 10^{-22}, by reaching deeper into the universe, would detect several events of this type per year.

The above examples illustrate impulsive or burst sources; some periodic sources have also been posited. For these the anticipated gravitational-wave strains are much smaller; and correspondingly any practical search for them will most likely be restricted to our galaxy. A compensation, however, is that the observations can be extended over long integration times to improve strain sensitivity. Pulsars (rotating neutron stars) would emit gravitational radiation as a result of any deviations from axial symmetry; the radiation frequency can be at the pulsar rotation frequency and at twice that frequency. The gravitational wave's strain amplitude is proportional to the ellipticity of the source. If the Crab or Vela pulsars had ellipticities as large as 10^{-5}, they would produce periodic strains at the Earth of 10^{-26} at 60 and 22 Hz, respectively. These strain amplitudes could be within reach of some proposed detectors after a month of integration (see Figure 6.3 in Chapter 6).

A final category of cosmic gravitational radiation is the stochastic background—a gravitational-wave background noise detectable as a correlated noise component in the output of a pair (or more) of detectors. The sources of such a background would most likely reside in the early universe, probably at epochs not accessible by electromagnetic radiation. Since a gravitational-wave background has energy density, experimental limits are usually quoted in terms of the universe's closure density ρ_c.*

---

* Closure (or critical) density ρ_c is that density that results in sufficient gravitational force eventually to stop the universal expansion. Currently, $\rho_c \approx 10^{-29}$ g cm^{-3}.

This partial listing of hypothesized sources has focused primarily on phenomena that might produce radiation at high frequencies, say 1 Hz to 10 kHz—the spectral band accessible to detectors on the ground. At lower frequencies from 1 μHz to 1 Hz, space techniques and astrophysical observations must be used to search for gravitational waves. Probable sources include classical binary star systems and white-dwarf binary systems in the 10^{-1} to 10^{-5} Hz region with strain amplitudes of roughly 10^{-22} to 10^{-20} and bursts associated with the formation and dynamics of massive black holes. This band contains the only astrophysical sources of gravitational radiation whose properties are well known—the nearby binary stellar systems. One particularly favorable source is ι Boo, a nearby binary system that produces a strain amplitude of around 10^{-20} at a period of 193 minutes. PSR 1913+16 is a disappointing source for direct detection because of its large distance from the Sun. The expected strain amplitude at multiples of the orbital frequency (10^{-4} Hz) is of the order of 10^{-23}.

DETECTORS

The first gravitational-wave detectors intended to sense waves of cosmic origin were demonstrated in the late 1960s. These detectors were aluminum cylinders instrumented to detect excitations of the bar's fundamental quadrupole mode by passing gravitational waves. The bars, typically of 1-ton mass, were suspended in vacuum chambers on shock mounts to reduce acoustic and seismic noise. They were operated at room temperature and achieved sensitivities limited only by thermal excitation of the quadrupole mode, a remarkably small noise amplitude. Coincidence detection with two separated bars was used to reduce accidental events. Experiments with such Weber bars have continued in several research groups throughout the world. Instrumentation improvements and cooling of the bars have helped to achieve a recent major improvement in sensitivity.

The second main class of detectors, the laser interferometers, began development later and is less mature. In these detectors the change in propagation time of light traversing a gravitational wave is measured. The polarization of quadrupole (Einstein tensor) waves causes changes in the propagation time of light with opposite sign in orthogonal directions transverse to the direction of gravitational-wave propagation. Laser-interferometer detectors exploit this polarization property by measuring the time difference of light propagating along the orthogonal legs of an L-shaped interferometer whose mirrors are attached to three freely suspended masses. The time differences are measured interferometrically with high precision. The effect grows with the time of interaction between the light and the gravitational wave, so multipass cavities are used.

The laser detectors are currently less sensitive than bars, but a large increase in sensitivity is expected if long baselines can be achieved. A ground-based system with 5-km baselines is currently being proposed, and a Sun-orbiting interferometer with 10^6-km baselines has been envisioned.

Bar and interferometric detectors have been built and operated only on Earth, not in space. Earth-based operation carries with it the heavy penalty of seismic noise and noise due to the gravitational effects of nearby moving masses. Isolation from seismic noise at kilohertz frequencies is practical, but isolation becomes increasingly difficult at lower frequencies, with the eventual barrier lying probably in the range of 1-10 Hz. Therefore it is necessary to consider space-based detectors in order to search at lower frequencies.

One kind of space-based gravitational-wave detector has been achieved by tracking of interplanetary spacecraft. Here the gravitational-wave experiment is only one of several scientific experiments sharing the mission. Passing gravitational waves cause deviations in both the spacecraft trajectory and the trajectory of the Earth; the characteristic time signature of a gravitational wave in the two-way tracking system helps to discriminate it from other effects in the tracking data. Light travel time to interplanetary spacecraft is minutes or hours, so the experiment is most sensitive to gravitational waves with frequencies in the millihertz band.

Still another kind of detector is achieved by substituting a radio pulsar for the spacecraft. Here one has only one-way rather than two-way signals and is at the mercy of the stability of the pulsar pulse period and pulse shape. Nevertheless pulsar timing is currently providing the best way of searching for possible gravitational waves in the microhertz frequency range.

5

Search for Gravitational Waves: Highlights

BINARY PULSAR

General relativity predicts that a binary stellar system will lose energy in the form of gravitational waves, so that the orbital period will decrease as the two stars spiral together. Although many binary systems are known, only for the binary pulsar system PSR 1913+16 can the motion of the system be measured accurately enough to test this prediction. Moreover, most stellar systems do not provide clean tests of gravitational physics for point masses, because tidal interactions, changes of stellar mass distribution, and mass exchange or mass loss cause unpredictable and often large changes in the orbit. Fortunately the binary pulsar does seem to be clean according to available observational evidence (see section on Systems of Compact Stars in Chapter 3).

Observations of the orbit of the binary pulsar over the 10 years since its discovery have shown that the orbital period is decreasing at a fractional rate of $(2.71 \pm 0.10) \times 10^{-9}$ per year (see Figure 5.1). General relativity predicts an orbital decay rate due to gravitational-wave emission of $(2.715 \pm 0.002) \times 10^{-9}$ per year. This agreement is a most impressive and beautiful confirmation of the theory and provides strong evidence for the existence of gravitational waves. Still, one cannot completely rule out the unlikely possibility that tidal and/or mass exchange effects conspire to just compensate for an error in the

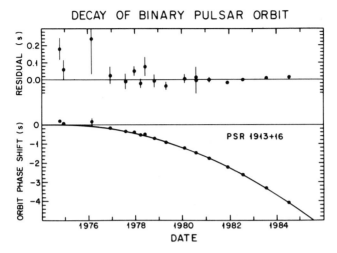

FIGURE 5.1 Evidence that gravitational radiation is correctly predicted by general relativity. The predicted change in orbit phase due to gravitational radiation by the binary system is shown by the solid line; dots are the observations, including errors. Residuals are shown with the expanded scale on the upper graph. The orbital motion (period ~8 hours) modulates the phase and frequency of the pulsar. By following the pulsar phase for many years, the orbit is measured with exquisite accuracy.

rate predicted by general relativity. Independent evidence that the pulsar's companion star is also a collapsed star would settle this issue.

In any case, these results already place stringent restrictions on alternative theories of gravity; in many theories, the decay rate of binary systems containing neutron stars or black holes is much greater than in general relativity theory owing to dipole gravitational radiation. In general relativity, monopole and dipole radiation are absolutely forbidden, and the lowest allowed mode is quadrupole radiation. The orbital decay observed in the binary pulsar is completely consistent with the quadrupole formula of general relativity.

BAR DETECTORS

Bar detectors have undergone 20 years of development, resulting in improvement of strain sensitivity by more than 4 orders of magnitude (8 orders of magnitude in energy-flux sensitivity). Major improvements achieved in the past decade include the following: cryogenic cooling; increase of the Q of bar materials to values approaching 10^8 in aluminum and exceeding 10^9 in sapphire and silicon monocrystals; improvements in several transducer types including inductive, capacitive, and resonant cavities; and improvements in coupling schemes and amplifiers. Vigorous work is continuing on all these critical and generally useful technologies.

Recently a bar antenna (see Figure 5.2) has been operated for several months at pulse-strain sensitivities of about 10^{-18} in a narrow-band mode near 1 kHz. No gravitational-wave signals were identified, but the thermal noise limit for the 4-K bar was achieved. Operation in coincidence of two or more bar detectors, distant from one another, permits much better detection capability by eliminating noise and interference events generated locally. Such coincidence observations have only been carried out over short time periods with recent detectors, although they were made over long intervals with early bar detectors.

No fundamental barriers are apparent to further improvements in the sensitivity of bar detectors by several orders of magnitude. Moreover, several current instrumentation developments could significantly extend the bandwidth of bar detectors. When bar detectors reach a strain sensitivity of about 10^{-20}, they will approach the so-called naive quantum limit. This means that gravitational-wave excitations of the fundamental mode of an initially unexcited bar will amount to about one quantum of acoustic oscillation, and issues of quantum measurement of the bar's state will become crucial. Techniques are now known which in principle allow one to measure an arbitrarily small fraction of a quantum of excitation. These are known as quantum-nondemolition or backaction-evasion techniques, and work is now under way to develop them in practice. When other sources of noise are reduced so much that bars are at the naive quantum limit, these techniques will be needed.

INTERFEROMETRIC DETECTORS

Laboratory-scale interferometric antennas with arm lengths extending from 1.5 to 40 m are now in operation at several laboratories around the world. Two of these instruments have achieved displacement noise spectral densities of 10^{-15} cm $Hz^{-1/2}$ in the 1- to 10-kHz frequency range. The corresponding root-mean-square strain sensitivity over the 30- and 40-m baselines is 10^{-17} for a 1-kHz bandwidth.* One of these

---

* Strain spectral density $h(f)$ [$Hz^{-1/2}$] is used to characterize broadband radiation and detectors with wideband frequency response. For signals of finite bandwidth B, the strain is $h = h(f)B^{1/2}$. For example, a bar detector with $h = 10^{-18}$ has sensitivity $h(f) = 3 \times 10^{-20}$ $Hz^{-1/2}$ to a 10^{-3}-s impulsive signal.

FIGURE 5.2 A bar-type gravity-wave detector. The 5000-kg aluminum bar is shown end-on with its transducer mount and lead vibration filters attached. Also shown are the suspending wires, the cryostat, and the towers containing seismic isolation filters. This bar has been successfully operated at 4 K.

detectors uses about 200 mW of laser power and 100 beam passes in each arm, corresponding to a light storage time of 10 μs. The other detector uses several milliwatts of laser power and high-Q Fabry-Perot cavities to achieve a storage time of about 1 ms. At high signal frequencies the sensitivity of interferometric detectors is limited by the available laser power.

The principal technical efforts to improve detector performance are in two areas. The first is to enhance the displacement sensitivity by increasing the laser power in the interferometer while controlling the effects of scattered light. The power can be increased by using more powerful lasers and/or by recycling the light from the output port of the interferometer back to the input. The second major effort is to reduce the influence of random forces on the interferometer masses. The development of improved suspensions to reduce thermal noise and coupling to external acoustic and seismic noise is actively being pursued and is required in order to achieve adequate detector performance at low frequencies.

An important feature of interferometer antennas is that they are inherently broadband and can detect and measure the wave forms of all classes of sources: impulsive, periodic (even if the period is not known in advance), and the stochastic background. However, an interesting new concept would enable the antenna to be tuned to a possible source of known period and phase, for example a fast pulsar. The light beams in the two arms would be exchanged in synchronism with the source, thus accumulating signal while averaging out noise.

PULSAR TIMING AND MILLISECOND PULSARS

The observed slowing-down rates of a number of radio pulsars are stable enough to afford useful upper limits on the amplitudes of low-frequency gravitational waves. Gravitational waves would shake the Earth or the pulsar and cause deviations in the observed uniformity of the period drift rate.

Until 1982 the fastest known pulsar was the Crab nebula pulsar with a period of 33 ms. Then a radio pulsar, known as PSR 1937+214, with a period of only 1.6 ms (a rotational frequency of 642 Hz) was discovered during investigation of a known peculiar radio source. The slowing-down rate for this object has unprecedented stability for a pulsar; indeed, over time intervals longer than a few months it seems to have as stable a drift rate as any known clock, natural or man-made. The best-known limits on gravitational waves in the microhertz frequency range come from observations of this pulsar; already it has been shown that waves in this band cannot contribute more than 5×10^{-4} of the critical mass density of the universe (see Figure 6.4 in Chapter 6).

SOURCES OF GRAVITATIONAL WAVES—RECENT DEVELOPMENTS

Earlier we used nonspherical collapse of a stellar core as one example of an impulsive source of gravitational waves. However, current theoretical models of Type II supernovae manage to agree roughly with the observations by assuming that the core is spherically symmetric during collapse. Thus there is no good reason to believe that Type II supernovae are strong sources of gravitational waves. Type I supernovae are less well understood, and a consensus model does not exist, although many believe that short-period binary systems are involved. Some models of Type I events predict strong gravity-wave emission; others do not. For instance, one model posits a close pair of white-dwarf stars as the presupernova object; mass accretion causes one star to spin up and eventually collapse, perhaps to a neutron star. Such a binary system would be a strong source of gravitational radiation at frequencies below 1 Hz; and the stellar collapse would be highly nonspherical, producing a strong burst of gravitational waves with frequencies around 1 kHz. The properties of collapsing, rotating stellar cores are now the subject of active investigation, often involving large-scale numerical work.

Discovery of the binary pulsar, which probably consists of two neutron stars, emphasized the possibility that decaying compact/compact binary systems are strong sources. Discovery of millisecond pulsars showed that rapidly rotating neutron stars do exist. If born rapidly rotating, these cores could have been moderately strong sources of gravitational-wave bursts. If, on the other hand, they owe their fast rotation to subsequent spinup by mass exchange with a close companion, they could have been sources of periodic gravitational radiation. (This model assumes that they have been spun up above the threshold for secular instability for gravitational-wave emission.) It should be noted that only a few years ago, before these discoveries, most theorists saw little hope that neutron stars could be sources of detectable gravitational waves. Again nature has outrun our imaginations, emphasizing the need for sensitive measurements.

More conjectural sources might exist at millihertz and microhertz frequencies. These include collisions of massive or supermassive black holes, which may exist in galactic nuclei, and even primordial gravitational waves from an early inflationary era of the universe's expansion, or waves emitted by decaying cosmic strings, which, according to certain grand unified theories, would have been created by phase transitions in the early universe. Perhaps detection of their gravitational waves will be our best handle on these intriguing processes.

6

Search for Gravitational Waves: Opportunities

LASER INTERFEROMETER DETECTOR WITH 5-KILOMETER BASELINE

In the limit where random forces on the end masses dominate the antenna noise budget, the gravitational-wave amplitude sensitivity of a laser interferometer improves with arm length as $h \propto L^{-1}$ (assuming that L is less than half the wavelength). Existing laser interferometric antennas ($L \leq 40$ m) are usually limited by random forces at low frequencies and, as laser power is increased, may become so at many frequencies of interest in the gravitational-wave search. The way to overcome this noise limit on the sensitivity of interferometric antennas is to increase the arm length. A current study for the National Science Foundation envisions an interferometer with arm lengths of 5 km, increasing the strain sensitivity by factors of 10^2 to 10^3 over that of current interferometer antennas. (See Figure 6.1.) A further increase in laser power by a factor of 10^3 or 10^4 (10 mW to 10 or 100 W) will be necessary to bring the gravitational-wave search using 5-km baseline interferometric antennas into the sensitivity regime required to intersect the present estimates of source strengths. Figures 6.2-6.4 show the sensitivity prospects for 5-km baseline interferometric antennas along with estimates of strains due to impulsive, periodic, and stochastic gravitational-wave sources. Detection of several source types is anticipated.

50 GRAVITATION

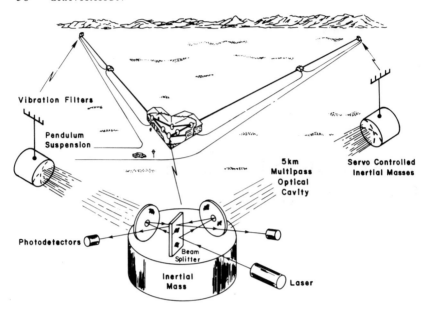

FIGURE 6.1 One half of a proposed long-baseline interferometric gravity-wave detector. A passing gravitational wave changes the light travel times differently in the two interferometer arms, causing a tiny shift in the light intensities at the detectors. The current design calls for 5-km vacuum pipes connecting the end stations. The signals from two such instruments, widely separated, are correlated to identify and remove effects from local noise sources.

Two stages of development are shown in the figures. The upper (solid) curve is the anticipated performance of current receiver designs in the large-baseline interferometric system. These receivers use modest extensions of the technology employed in the present prototypes. The lower (dashed) curve is the anticipated performance of second-generation receivers. Receivers of this sensitivity have been conceptually designed but not yet constructed and will not be effectively tested until a large-baseline facility is available. To emphasize the importance of increased sensitivity we note that, if extragalactic sources can be reached (e.g., decaying neutron star binary systems in the Virgo cluster), the event rate increases dramatically, scaling as h^{-3}, which varies with arm length as L^3 for interferometers limited by certain types of noise.

It is expected that increases in laser power and seismic isolation will not require great technical advances or expense. The main expense of long-baseline interferometers is in the vacuum system and site con-

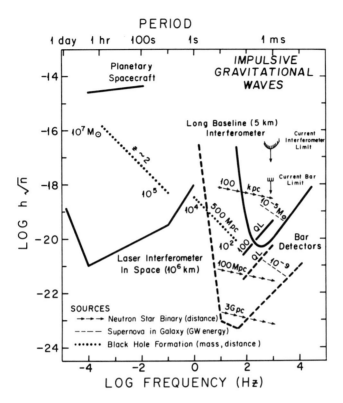

FIGURE 6.2 Prospects for detecting impulsive gravitational waves. This figure indicates projected sensitivities of the various gravitational-wave detection schemes for impulsive or burst sources. The sensitivities are given in terms of the rms strain noise of the detectors in a frequency band equal to the reciprocal pulse length. In order to compare them with the strain amplitude of the hypothetical sources shown, the detector sensitivities should be degraded by a factor of $[\ln(T_p R)]^{1/2}$, where T_p is the pulse length and R the event rate, to account for pulse detection statistics. Binary-system decay events are quasi-periodic; the detection sensitivity for these events improves as the square root of the number of cycles n observed in the wave train. Two assumptions are made for the ground-based detectors. The solid curves show the sensitivities possible with modest extensions of current technology; the dashed curves assume some advanced development. For example, in the 5-km interferometer it is assumed that initially the optical power will be 10 W with a light storage time of 1/2 (gravity-wave period); the dashed curve assumes that laser power is 100 W, mirror reflectivity is 0.9999, light is recycled from the output port back into the input port, and seismic noise will be eliminated for $f > 10$ Hz. The projected bar detector consists of an array of four resonant masses ranging in mass from 5×10^3 kg (840 Hz) to 42×10^3 kg (100 Hz). The dashed curve assumes a quantum-limited (QL) linear amplifier; the solid curve assumes an amplifier with 100 times more noise.

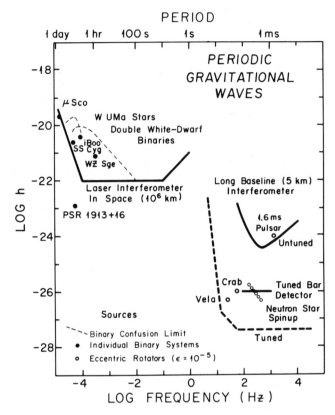

FIGURE 6.3 Prospects for detecting periodic gravitational waves. This figure is drawn for detector integration times of 10^6 seconds (sensitivity improves as $\sqrt{t}$). The only guaranteed sources—the known fast binary star systems (e.g., ι Boo)—could be seen by the laser interferometer in solar orbit. In fact, the broad beams of this antenna would include many sources of measurable strength, and sensitivity may ultimately be limited by a background of weak sources. For increased sensitivity the ground-based antennas can be tuned to sources of known frequency, such as pulsars. The interferometer is tuned by synchronously exchanging the light beams between the two arms. A different resonant bar is needed for each source, but a single large cryostat could be used. The bar curve assumes $Q = 10^7$, $T = 50$ mK, and $m = 5 \times 10^3$ kg.

struction. Two antennas are envisioned to perform coincidence measurements, thus eliminating local-noise events.

BAR DETECTOR SENSITIVITY AND BANDWIDTH

There is currently a multifaceted development program in bar detectors, which promises to continue to improve the sensitivity and

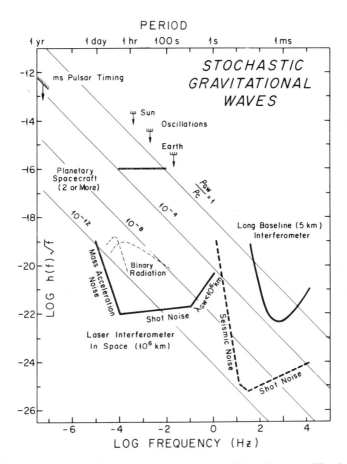

FIGURE 6.4 Prospects for detecting stochastic gravitational waves. The detector sensitivities in the figure assume that cross correlation of two antennas is carried out for an integration time of 10^6 seconds and the detection bandwidths are equal to the frequency, except for the dashed long-baseline curve where the bandwidth is narrowed by a factor of 10 owing to resonant interchange of light between the interferometer arms. The sensitivity improves as the product of the bandwidth and integration time to the 1/4 power. The straight lines in the figure are the strain spectral densities of a universe filled with the indicated fraction of the closure density in gravitational waves on the assumption that all the gravitational radiation power is concentrated in a bandwidth equal to the frequency. This figure also indicates those sources of noise that are expected to limit the sensitivity of the interferometric detectors.

bandwidth of searches for kilohertz gravitational waves. The next few years should see coincidence experiments carried out at a strain sensitivity better than 10^{-18}, at frequencies near 1 kHz, with bandwidths in the range 10-100 Hz.

Within a decade, further improvement of strain sensitivity by 2 to 3 orders of magnitude should be achievable, through further cryogenic cooling and use of advanced transducers and amplifiers. Techniques are also under study to increase the bandwidth of bar detectors; the use of cascaded, strongly coupled mechanical resonators can in principle give both high sensitivity and wide bandwidth in a single bar. It is estimated that bandwidths of several hundred hertz or more can be achieved.

Development of low-noise amplifiers that can be well coupled to transducers is important, notably superconducting quantum interference devices (SQUIDs). In the past, the gravitational-wave community has mostly depended on outsiders to develop improved SQUIDs; continued support for SQUID development is an important element in the bar detector program.

Within the next decade, operational bars could approach the naive quantum limit. Techniques for passing beyond the limit will be necessary then, and current ideas merit study.

Arrays of bars also provide a path to greater sensitivity and wide bandwidth. The strain sensitivity of an array increases as the square root of the number of bars; bandwidth can be increased by tuning different bars to different frequencies—a "xylophone" for gravitational waves.

OBSERVATIONS WITH BAR DETECTORS

At current sensitivity levels, the event rate for burst sources of gravitational waves is thought to be only about 1 per 10 years or worse. Therefore, detector development and construction should take precedence over major observing programs at present. However, some significant observing runs are desirable for two reasons: to keep development attuned to the actual problems that occur in observing, and especially to understand any noise or interference that appears; and not to miss a chance to see sources should the theoretical best estimates be quite wrong. We again emphasize that nature has provided stronger sources than theorists predicted in the electromagnetic radiation bands. Coincidence observations should be planned for every order-of-magnitude enhancement in strain sensitivity.

PULSAR SEARCHES

The discovery of a binary pulsar in a clean system, and also the later discovery of several pulsars with periods in the millisecond range, have been of great importance for the study of gravitational radiation. Further progress could come with more such discoveries; for example, the availability of several pulsars with the short period and excellent frequency stability of PSR 1937+214 would in principle allow, by cross-correlation, a sensitive search for gravitational waves of microhertz frequency passing through the solar system. A deep radio search for fast pulsars should have high priority. Such a search will require a substantial investment in data processing, both on-line and off-line. Further searches for millisecond periods among x-ray pulsars should also be carried out, because accreting neutron stars with rotational periods in the millisecond range could be significant periodic sources of gravitational waves.

SPACECRAFT TRACKING

Accurate tracking of interplanetary spacecraft offers, at present, our only opportunity to search for gravitational radiation in the frequency range 10^{-2} to 10^{-4} Hz. The long travel time of interplanetary signals and the inherent precision of time measurement account for the good sensitivity of this technique to low-frequency waves. With a single spacecraft the method is most sensitive to impulsive gravitational waves, but the use of two spacecraft makes possible a search for a stochastic background as well. Sensitivity estimates are shown in Figures 6.2 and 6.4.

Preparations are currently being made to search for gravitational waves using the Galileo mission to Jupiter and the Ulysses (formerly International Solar Polar) spacecraft. About 40 days of observations are planned to start in October 1987 when both spacecraft are near Jupiter. For Galileo, the National Aeronautics and Space Administration has arranged for X-band tracking on the uplink and S- and X-band frequencies on the downlink. The expected system sensitivity to impulsive radiation with frequency components in the 10^{-4}- to 10^{-2}-Hz range is $h = 3 \times 10^{-15}$—a factor of 10 improvement over past spacecraft. To improve substantially the sensitivity for gravitational-wave detection beyond the level expected for Galileo, two main types of noise source must be addressed. Fluctuations in the interplanetary and ionospheric electron densities can be measured and removed by using two tracking frequencies on the uplink and the downlink. The effects of variable tropospheric delay can be reduced by atmospheric monitors or, better yet, effectively eliminated by using signals from a high-stability clock on board the spacecraft.

Clearly, it is important to consider these needs early in the planning stage of a spacecraft mission if sensitivity to gravitational waves is to be optimized. The impacts on mission configuration and cost are relatively small.

SPACE INTERFEROMETERS

Any earthbound gravitational-wave detector is subject to seismic noise, which in practice imposes a lower cutoff on the detectable gravitational-wave frequency, in the neighborhood of 1 Hz. For high sensitivity at lower frequencies (10^{-6} to 1 Hz) a laser interferometer in space is an attractive possibility. Separate spacecraft would carry the three interferometer end stations, as shown in Figure 6.5. Preliminary studies envision the three spacecraft orbiting in formation around the Sun, with 1-year periods and with separations of about 10^6 km. Lasers of 1-mW power in each station would communicate using 50-cm-diameter mirrors, with the end station lasers phase locked to the signals received from the central station. The end mirrors and central beam splitter for the interferometer are mounted on masses that are protected from spurious forces due to the solar wind and solar radiation pressure. For this system the anticipated sensitivity is $h \approx 10^{-22}$ for narrow-band periodic signals and 10^{-19} to 10^{-20} for pulses at frequencies of 10^{-4} to 10^{-1} Hz. The sensitivity degrades outside this range but is still useful from about 10^{-6} to 1 Hz, as seen in Figure 6.3.

This sensitivity would allow detection of the known nearby binary system ι Boo, if it is radiating as predicted by general relativity. This is also the frequency range for detecting broad spectral features due to the superimposed radiation from many white-dwarf binary systems and from classical binary systems. The expected energy density in gravitational waves from such sources is about $10^{-8} \rho_c$ (see Figure 6.4). The conjectured massive black holes would also radiate in the millihertz band. Detectable pulses of gravitational radiation are possible from pregalactic or early galactic formation of massive black holes, from coalescence of such objects, or from their falling into other massive black holes that may exist at the centers of galaxies. Observation of such events would have far-reaching consequences for gravitation, astrophysics, and cosmology.

More advanced studies of a Sun-orbiting laser interferometer system

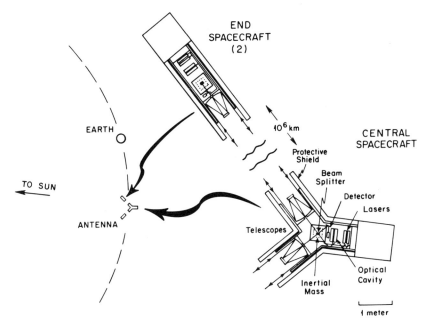

FIGURE 6.5 A concept for a gravity-wave detector in space. The basic principles are the same as for the 5-km ground-based detector (Figure 6.1). However, the longer baseline and freedom from seismic noise permit operation at low frequencies—1 Hz to 10^{-6} Hz. Passive optical cavities would be used for precise frequency control, and the mirror-carrying masses would be shielded from solar-wind buffeting.

are needed to evaluate the technical and cost aspects of a possible space mission.

EVENT RATES AND SOURCE CALCULATIONS

Theoretical activity in modeling possible sources, and in attempting to determine their frequency of occurrence in the universe, is key to an effective search for gravitational waves. The main uncertainties in theoretical estimates of gravitational-wave source properties are not due to physical understanding, which we think is good, or computational ability, which is already considerable and steadily improving, but to our uncertainties about the astrophysical boundary conditions. Easy answers are not to be expected, and the best support for the experimental program comes with the investigation of all plausible sources and the best possible estimates of their observable properties.

COMPUTATION

The computation requirements for operating gravitational-wave detectors have not been studied in detail. In one mode of operation, namely in searches for narrow-band periodic sources of unknown frequency and celestial position, the computational needs are likely to be large but not impossible. The difficulty of the data-reduction problem in this mode of operation arises because it is necessary to search simultaneously in three parameters (frequency, right ascension, and declination). Therefore, as multidetector observations get under way, appropriate computing facilities will be needed.

Deep radio pulsar searches at millisecond periods, and searches for millisecond periodicities in known x-ray sources, also require substantial computational power.

Computational needs for gravitational-wave source calculations are discussed in Chapter 9 in the section on Computation.

7

Gravitation Theory: Introduction

Few areas of physics apply to such a broad range of phenomena as does Einstein's General Theory of Relativity. Gravity, which it describes, governs our universe on its largest scales and its smallest. Current attempts to construct a quantum theory of gravity confront physics at its most fundamental level. At the same time, relativity is an important element in the variety of physics being used to construct models of pulsars, x-ray sources, quasars, and the universe itself. The theory of relativity is deep and central and has broad application. As a consequence it is actively worked on and has many ties to other disciplines.

The General Theory of Relativity was proposed in final form by Einstein in 1916. It is at once a theory of gravity and a theory of the structure of space-time; it attributes the gravitational interaction to the geometric curvature of our space-time continuum. The late 1960s saw the start of a period of exciting development in the area of relativity. In part this was brought on by new astronomical discoveries—the discoveries of pulsars, galactic x-ray sources, possibly black holes, and the 3-K cosmic radiation. Further stimulation came from the fruitful interaction of particle physics and quantum gravity in the development of field-theory techniques that could be applied to both areas and the progress in particle physics toward energy scales where quantum gravity must be important. Fundamental discoveries within the field (the singularity theorems and the Hawking radiation to give just two examples) deepened our understanding, provided new confidence, and enabled the theory to be extended into new domains. This exciting period of development is still under way.

Gravity is a very weak force by the standards of elementary-particle physics. The only reason that it is the dominant force on astrophysical length scales is that it is always attractive; unlike electromagnetism, it cannot be canceled, neutralized, or shielded against. Even in our solar system, gravity is a weak force by relativistic standards—matter velocities are everywhere less than the speed of light by a factor of 10^{-3}. The crucial effects that mark the difference between Newtonian theory and general relativistic gravity and between general relativity and alternative relativity theories are therefore very small in our solar system. Nevertheless, there are times and places in the universe where gravity is strong, such as in neutron stars, black holes, and the big bang. Even the universe itself is a highly relativistic system in that recession velocities approach the speed of light for the most distant known objects.

The General Theory of Relativity, as a fundamental theory of physical interactions, contains three major kinds of purely gravitational elementary objects: gravitational waves, black holes, and isolated universes. The simplest universe is the flat, empty universe of Minkowski space-time. Other simple universes are the homogeneous, isotropic cosmological models, which are good models for our universe. These elementary objects can be combined in various ways: Gravitational waves can propagate, either on a flat background or in a universe. Black holes can also inhabit either. Black holes can be formed by the implosion of sufficiently intense pulses of gravitational waves, as well as by the gravitational collapse of matter. When black holes collide, they coalesce to form a larger black hole, and some gravitational waves are radiated.

When the effects of quantum mechanics are included, new processes appear. Black holes decay by quantum emission of particles and eventually disappear (Hawking radiation). Universes can in principle tunnel into one another by quantum-mechanical barrier penetration. Quantum effects are important for the structure of a big-bang singularity.

8

Gravitation Theory: Highlights

NEUTRON STARS

The discovery of radio pulsars in 1967, and their subsequent identification as rotating neutron stars, exhibited for the first time objects in the universe with gravity so strong that the effects of general relativity must be important in their structure. Subsequently, neutron stars were also discovered in x-ray binary systems, in which gas from a normal star is accreting onto the neutron star and releasing gravitational binding energy as x rays. The discovery of the binary radio pulsar PSR 1913+16 has provided an impressive example of relativistic effects in its orbital motion (see earlier sections on Perihelion Advance, Einstein's Only Handle in Chapter 2 and on Binary Pulsar in Chapter 5). However, it has been difficult to discover direct evidence for general relativistic effects in observations of neutron stars themselves. One likely case is the observation of a spectral line in hard x-ray observations of gamma-ray bursters, which can be understood as the positron annihilation line emitted at the surface of a neutron star, redshifted about 10 percent by the general relativistic gravitational redshift.

The theory of stellar structure and stellar pulsation, which was originally developed for normal stars, has now been successfully extended to neutron stars, taking full account of general relativity. The most important new effects are the emission of gravitational waves and the existence of relativistic instabilities (not present in Newtonian theory), which limit the physical range of stellar possibilities. Detailed results are available for the frequencies and damping rates for nonradial oscillations of neutron stars. These oscillations generate gravitational-wave emission and might be excited during the birth of a neutron star in stellar collapse. Surprising new effects have been uncovered for gravitational-wave emission by rotating stars. A general theorem has been proved that says roughly that all perfect fluid, rotating stars are unstable to the emission of gravitational radiation via a secular instability. The inclusion of viscosity, always present to some degree in nature, allows the instability to exist only above some critical threshold in stellar rotation rate. The recent discovery of a pulsar with a rotational period of only 1.5 ms has shown the relevance of these results, and it now seems quite possible that there exists a class of fast pulsars with rotational rates at or near the instability threshold, which could be sources of periodic gravitational radiation.

GRAVITATIONAL COLLAPSE AND BLACK HOLES

There is a maximum mass limit for neutron stars. The exact limit depends on the equation of state for nuclear matter, which is not well known. Nevertheless, the upper mass limit is certainly less than about 5 solar masses and seems likely to be in the range of 1.5-2.5 solar masses. Any stellar core with a mass exceeding the upper limit that undergoes gravitational collapse must collapse to indefinitely high central density to form a singularity. It is generally believed among theorists that in such circumstances a black hole will always form so that the final singularity will be hidden from external observers. A black hole is a region of space-time where the gravitational field is so strong that not even light can escape. Inside a black hole, there exists a space-time singularity, a place where the space-time curvature becomes infinite and all the known laws of physics may break down. The hypothesis that space-time singularities always remain hidden from observers is called the cosmic censorship hypothesis. It remains unproved. A singularity that is visible to external observers, in violation of the hypothesis, would be naked singularity. Naked singularities have been the object of a considerable amount of study and speculation, but most relativists believe that they do not exist, with the exception of the big bang itself.

The evidence for existence of black holes is impressive but not conclusive. At least one binary stellar x-ray source, Cygnus X-1, has a mass greater than the upper mass limit for neutron stars and must be of compact size as evidenced by millisecond variability of its emission. Alternative models are possible but implausible. Other x-ray sources may well contain black holes; a strong possibility is LMC-X3 in the Large Magellanic Cloud.

Supermassive black holes of a million to a billion solar masses might be able to form in the nuclei of galaxies by stellar coalescence and accretion. Direct evidence for such black holes remains weak. Although some galactic nuclei are found to possess a large accumulation of dark matter at the core, this mass has not been shown to be so compact that it must be a black hole. A popular model of quasars and active galactic nuclei postulates the existence of a supermassive black hole undergoing accretion of surrounding matter, with enormous amounts of gravitational energy released in thermal and nonthermal radiation coming from an accretion disk or a chaotic accretion region around the black hole. Accretion may produce electrodynamic effects or jets of outflowing matter.

A black hole may be born in a more or less excited state, depending on the degree of disorder in the gravitational collapse of its progenitor star, but it quickly relaxes to a stationary state by the emission of gravitational waves. The stationary states of black holes are remarkably simple, according to the uniqueness theorems, which state that a stationary black hole must belong to the three-parameter family of black-hole solutions of the Einstein equations, the Schwarzschild-Kerr-Newman solutions. The three parameters are the mass, the total angular momentum, and the electric charge. The mass, angular momentum, and charge of a black hole cannot disappear because these quantities are conserved charges coupled to long-range fields.

In an astrophysical environment, any electric charge on a black hole will be quickly and almost completely neutralized through the conductivity of the surrounding plasma. Since the electromagnetic interaction is so much stronger than gravity, it only takes a tiny amount of charged plasma to neutralize even a maximally charged black hole. On the other hand, the angular momentum of a black hole will persist, and a black hole may remain rotating for hundreds of millions of years or more.

A rotating black hole has free energy that can be tapped externally. The energy can be tapped by immersing the black hole in a suitable configuration of conductors and magnetic fields, in which case it acts as a kind of electrical generator, or it can be tapped mechanically by suitable arrangements of particles traversing a certain region near the black hole, called the ergosphere. When all the free energy is removed from a rotating black hole, it is reduced to a nonrotating state.

The energetics of black holes are governed by remarkably simple laws, the four laws of black-hole dynamics, which in essence are the four laws of thermodynamics as applied to black holes. The role of entropy is played by a purely geometric quantity, the surface area of the black hole. The area theorem states that, in classical physics, the surface area of a black hole never decreases; this theorem is known as the second law of black-hole dynamics from its parallelism to the Second Law of Thermodynamics, which asserts that entropy never decreases for an isolated system.

Small disturbances of stationary black holes, for instance due to small particles falling in or to impinging electromagnetic or gravitational waves, can be worked out in linear perturbation theory. This theory reduces to the solution of certain remarkably simple wave equations and is in essence complete.

QUANTUM PARTICLE CREATION BY BLACK HOLES

A major advance in fundamental understanding of the laws of physics was achieved in the discovery that, when quantum effects are considered, a black hole emits quanta of radiation just as if it were a blackbody at a finite temperature. The temperature is inversely proportional to the mass of the black hole. The temperature of a black hole is its key feature that makes possible the identification of the laws of black-hole dynamics with the laws of thermodynamics. The emission of quanta by black holes, known as the Hawking process, causes black holes to become gradually smaller and finally to decay away entirely. This effect seems unobservable for stellar mass black holes, for which the temperature is less than a microkelvin and whose lifetime is greater than 10^{70} years. On the other hand, if small black holes, with a mass of about 10^{15} g, were created in the big bang, they could be observable today. Their lifetime would be roughly 20 billion years, the age of the universe, and their temperatures would become high just before their final decay, so that they would emit a burst of hard electromagnetic radiation. Such bursts have not been found to date.

QUANTUM EFFECTS IN THE EARLY UNIVERSE

Gravity becomes comparable in strength with the other fundamental forces of nature only at the Planck energy, about 10^{19} GeV. The only known places in the present universe where energies reach this level are at space-time singularities. Those inside black holes are thought to be invisible to us, according to the cosmic censorship hypothesis. The initial singularity of the big bang is in principle observable to us, although it is shrouded in the hot dense matter of the primeval fireball. The effects of quantum gravity, imprinted on the universe at times so early that the temperature exceeded the Planck temperature, could have affected the present universe in important ways. An important effect could have been the damping of initial anisotropies, by quantum particle creation, to leave the almost perfectly isotropic universe that we see today. On the other hand, residual anisotropies in the cosmic background radiation could have been created by quantum effects at somewhat later times, for instance during an inflationary era in cosmology (see the section on The Inflationary Universe in Chapter 12).

ALTERNATIVE THEORIES

The tremendous advances in experimental tests of relativity have changed the theoretical scene greatly in the last decade. Theories that were viable and indeed admirable have now been stringently constrained or even ruled out by solar-system tests and by observations of the binary pulsar PSR 1913+16. This progress has increased the confidence of most gravitation theorists that general relativity is indeed the correct classical theory of gravity, at least in the long-distance, low-energy domain, despite the fact that many of its most important effects, such as detection of gravitational radiation and magnetic gravity, remain to be demonstrated. Thus, although some work continues on alternative theories, most ongoing theoretical work is based on general relativity.

EXACT SOLUTIONS OF THE EINSTEIN EQUATIONS

The Einstein equations are a nonlinear set of coupled partial differential equations, and their complete solution is unknown. The discovery of exact particular solutions has played an important role in the progress of relativity; for instance, the Kerr solution, which is now known to be the unique solution for a rotating, uncharged, stationary black hole, was first found in a systematic search for certain exact solutions known as algebraically special.

Great progress has been made on solution of the Einstein equations in the more general case of a stationary, axisymmetric, vacuum space-time, which is now known to be completely soluble in principle. Soliton methods from mathematical physics have also been applied to this problem.

There has even been reason to hope for the complete and general solution of the Einstein equations. A set of ideas called twistor theory has been developed in a new approach to the issues both of classical and of quantum general relativity. Twistor theory has close connections to modern mathematics, specifically to algebraic topology and algebraic geometry. Twistor theory has already produced new exact solutions for non-Abelian gauge theories in field theory (some of the instanton solutions) and has also produced large new classes of complex valued solutions to the Einstein equations. There has been progress toward a general solution by twistor techniques, though as yet it has not been achieved.

The initial-value problem for the Einstein evolution equations is itself a deep problem, on which good progress has been made. Known exact solutions for the initial-value constraint equations are few, but constructive methods are now available that give, in principle, the general solution of the Cauchy (spacelike) initial-value problem from freely specifiable initial data for the gravitational field and matter fields. Characteristic (lightlike) initial-value surfaces are likewise often useful, especially in the study of gravitational radiation.

ASYMPTOTIC PROPERTIES OF SPACE-TIME

An isolated system in general relativity is represented by a space-time that becomes asymptotically flat (Minkowskian) at infinity. Physics should become simple at infinity; and mass, angular momentum, and gravitational waves should become easily measurable there. However, the nonlinearities in the Einstein equations make the study of infinity a subtle one.

In a general, asymptotically flat space-time in which gravitational waves are propagating toward infinity, the Riemann curvature tensor falls off only as $1/r$ in the directions (called null infinity) in which both t and r become large. This slow falloff of the curvature causes many difficulties in principle for the measurement of the properties of isolated systems by distant observers, for instance in the definition of angular momentum. It has been found that if one generalizes the space-time manifold into a four-complex-dimensional manifold by allowing the four space-time coordinates to take complex values instead of just real ones, then a reference system of remarkable simplicity exists at infinity. In it the most troublesome asymptotic terms in the geometry vanish. The new complex space that arises at null infinity is called H-space or the nonlinear graviton. Asymptotic properties of space-time at spacelike infinity (t fixed, r large) also reveal subtleties. The correct definition for the angular momentum of an isolated system as measured at spacelike infinity is a problem that has only recently been resolved.

NUMERICAL RELATIVITY

In the absence of analytic techniques for the general solution of the Einstein equations, relativists have turned to large-scale numerical techniques to solve important problems, such as the collapse of stellar cores or the collision of black holes. The inclusion of general relativity in spherically symmetric computations of stellar collapse is now routinely done when necessary. Nonspherical systems, which unlike spherical ones admit gravitational radiation, are much more difficult to simulate numerically and require both state-of-the-art numerical techniques and the largest computers. These computations also require state-of-the-art theoretical analyses of the Cauchy and characteristic initial-value problems of general relativity. Finally, great care is needed for the numerical treatment of hydrodynamics in these simulations.

The most ambitious numerical calculation carried out to date in pure general relativity, without any matter present, is the head-on collision of two identical nonrotating black holes. The numerical results show that the two holes coalesce to form a single one, and gravitational waves amounting to about a part in 10^3 of the total rest mass of the system are radiated to infinity.

EMISSION OF GRAVITATIONAL RADIATION

Inspired by experiments to detect gravitational radiation, investigators have studied many source models. The calculations carried out include perturbation studies of gravitational collapse and black holes, approximate models of collapsing cores and of colliding neutron stars, and full-scale numerical calculations of gravitational collapse, colliding neutron stars, and colliding black holes (see Figure 8.1). As noted in Chapter 2, the results of such calculations are essential for making the important estimates of the strengths and frequencies of gravitational waves near the Earth (Figures 6.2 and 6.3 in Chapter 6).

Doubts were raised about the validity of the quadrupole formula for gravitational-wave luminosity and radiation reaction of weak-field, slow-motion systems. Careful investigation of this formula by techniques of applied mathematics have strongly reinforced the belief in its validity. The experimental confirmation of the prediction of this formula for the binary pulsar PSR 1913+16 has emphasized its importance.

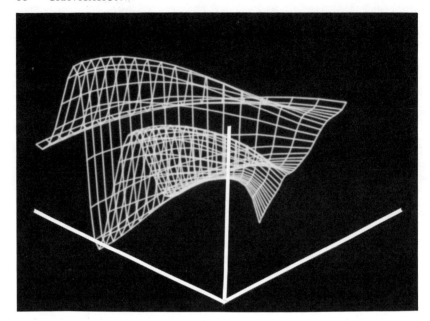

FIGURE 8.1 Numerical simulation of gravitational radiation from two colliding black holes of equal mass. The right axis is the axis of symmetry for the collision, and the left axis lies in the equator. Wave amplitude is plotted upward. This is the outgoing wave at a time of about $t = 37$ (mass) after the collision.

THE POSITIVE ENERGY THEOREM

Gravitational binding energy is negative, because gravity is an attractive force. When a body of given mass becomes so compact that the effects of general relativity become significant for its structure, the binding energy becomes comparable with the total rest energy of the matter making up the body. The possibility thus arises that the total energy of the body could become negative, should the binding energy actually dominate. It was conjectured 20 years ago that the total energy of a body could never become negative in the General Theory of Relativity. Heuristically one expects that any body attempting to violate this condition would lose stability and collapse to form a black hole before its total energy could become negative. A general form of this conjecture was finally proved in 1979 by two mathematicians using sophisticated arguments from differential geometry, and several general forms of this positive energy theorem have now been proved. Mathematicians were attracted to the problem after relativists publicized the importance and apparent difficulty of the conjecture.

A quite different and more direct proof of the positive energy theorem was given in 1981 by a particle physicist, using an argument motivated by supergravity theories (see section below on quantum gravity). Two relativists had earlier shown that the Hamiltonian of supergravity—the expectation value of which is the total energy—is formally nonnegative because it is a sum of perfect squares of certain fermionic charges. When this formal argument is made concrete, it indeed yields a rigorous proof of positive energy in general relativity.

QUANTUM FIELD THEORY IN CURVED SPACE-TIME

The discovery of the Hawking process by which black holes radiate particles quantum mechanically led to extensive development of the theory of quantum-matter fields in curved background space-time. A deeper understanding of the Hawking process was achieved together with a compelling and suggestive unification of the laws of black-hole mechanics and the laws of thermodynamics. The theory served as a laboratory in which ideas eventually to be important in a quantum theory of gravity could be tested in a simpler situation. Many conceptually interesting and unanticipated ideas emerged. The reaction of a moving particle detector to a curved space-time, the possibility of CPT nonconservation in quantum gravity, and the possibility of quantum-mechanical evolution from pure to mixed states are three examples.

QUANTUM GRAVITY

The last decade has seen a remarkable growth in the theoretical effort devoted to the construction of a quantum theory of gravity. The unification of gravity and quantum physics had always been understood to be a fundamental question. The activity of the past decade was much stimulated by new techniques arising from gauge theories that could be applied to answer new questions in quantum gravity and to the ever more active search in particle physics for a unified theory of all interactions, which must at the end include gravity.

The standard approach to field theory in the 1950s and 1960s was through the perturbation theory for scattering amplitudes. This is not always a sufficient tool in non-Abelian gauge theories such as quantum chromodynamics, nor will it suffice for gravity. On the one hand, gravitational scattering processes are too weak for observation. On the other, the perturbation theory for these processes has divergences too strong to be controlled by renormalization. New techniques or new ideas were necessary, and they emerged through a fruitful exchange with particle theory ("ghosts" for example, originated in studies of quantum gravity). Euclidean functional integrals, successful in other areas of field theory, were applied to formulate a quantum theory of gravity based on the Lagrangian of general relativity. When a Euclidean formulation is applied to field theories of flat space-time, it is just a different technique; however, when it is applied to gravity it yields a different quantum theory. Further, it yields the theory in a way in which it can be approximated semiclassically in regimes far from the domain of validity of perturbation theory. New questions could thus be asked, and novel results emerged. For example, in this theory pure states can evolve into mixed states in striking contrast to the usual situation in quantum mechanics.

There was also progress in the more traditional canonical approach to quantum gravity. Functional integral techniques clarified some of this approach's central problems, and promising new formulations of the canonical framework were worked out. The gravitational measure in the path integral, the existence of trace anomalies for the stress tensor, and solutions describing topological nontrivial configurations are just some examples. More recently, non-Abelian anomalies and the quantum breaking of coordinate invariance provide other striking illustrations involving gravity, gauge theories, and recent mathematics.

Relativists have tended to interpret quantum gravity in terms of the quantum version of Einstein's theory. General relativity works well in the classical long-range limit. It has also been shown to be the unique theory of gravity in this limit, on the basis of a few observational facts taken together with the properties of special relativistic quantum mechanics. It is not self-evident, however, that it is correct on the scales of 10^{-33} cm (10^{19} GeV) that characterize strong quantum gravitational phenomena. The 1970s and 1980s therefore saw the investigation of new theories that were generalizations of Einstein's theory and also some radically different approaches. The twin motivations for these new initiatives were the hopes that a new theory might be more tractable at small distances than Einstein's theory seems to be, and the need for a new theory to realize the goal of the unification of all interactions.

The developments of the past decade have seen a dramatic increase in the diversity of approaches to a quantum theory of gravity. Clearly, at present, a variety of approaches offers the best hope for a solution to this fundamental problem. One cannot help but be excited and impressed by the beauty and potential of these ideas.

One of the most significant developments of the past decade has been the emerging close relationship between particle physics and gravitation physics on this fundamental frontier. The search by particle physics for a unified theory has led to the problems of gravitational physics, and the search for a quantum gravity has led gravitational physics to field theory. Goals, techniques, and to some extent people are now shared between the cutting edges of these two areas.

Supergravity, induced gravity, higher derivative Lagrangians, twistor theory, geometric quantization, discrete gravity, Kaluza-Klein theories, and string and superstring theories are just some of the headings under which new theories of quantum gravity might be grouped. It would be inappropriate to review them all here. Each has its promise and successes, but none has succeeded. We shall mention just two approaches that are currently under intense study by particle and gravitation physicists.

Supergravity

Symmetries between fields of different spins and different statistics are the basis of supergravity theories, which promote this symmetry into a local gauge invariance. The gravitational field is symmetrically related to a larger collection of fields that describe all particles and all interactions. Supersymmetric theories have a number of remarkable properties, such as a less rapidly divergent perturbation theory than ordinary gravity. Despite the absence of immediate direct experimental tests (a situation that is rapidly improving, however), they have captured the imagination of many theorists as one of the few viable avenues leading toward a unification of the forces of nature.

Kaluza-Klein Theories

There appear to be only four dimensions to space-time, but in the framework of Kaluza-Klein theories appearances are deceiving. These generalizations of Einstein's theory envisage a world of many (e.g., ten or eleven) dimensions in which all but four are curled up so as to be unnoticeable on our macroscopic scales. In such theories the matter degrees of freedom are space-time degrees of freedom in the extra dimensions. Kaluza-Klein theories offer the hope of a purely geometric unification of gravity with other matter interactions and perhaps even the explanation of the four-dimensional character of our physical world.

9

Gravitation Theory: Opportunities

Theoretical research depends most importantly on its human resources. Theorists are much more able than experimentalists to redirect their research programs when important new opportunities appear. Consequently, the needs and the health of theory are best discussed in terms of the vitality and diversity of the research programs of individuals. Similarly, any list of the most important problems in theory must be descriptive rather than prescriptive. The research problems discussed here are selected from the menu of topics that theorists currently consider important.

CLASSICAL GRAVITATION, SINGULARITIES, ASYMPTOTIC STRUCTURE

Although we now seem to have a decent understanding of the basic physics of the General Theory of Relativity in the nonquantum regime, outstanding problems of great significance remain. The most important of these is the Cosmic Censorship Conjecture (see section in Chapter 8 on Gravitational Collapse and Black Holes). The proof of this conjecture would confirm the already widely accepted and applied theory of classical black-hole dynamics, while its overturn would throw black-hole dynamics into serious doubt.

A number of related issues about asymptotic properties of space-time remain to be settled, although there has been enormous progress in the last decade in this area. The measurement and even the definition of angular momentum at null infinity needs further clarification. It is impossible to give a local and covariant definition of energy density for the gravitational field, owing basically to the Principle of Equivalence, which says that space-time is everywhere locally flat. Nevertheless, significant progress has come in quasi-local definitions, in which one attempts to measure the total mass energy within a closed surface, and further development of these ideas will be useful. One conjectured extension of the Positive Energy Theorem still remains unproved, namely, that the total mass of an isolated system containing black holes must not only be positive but must exceed the sum of the irreducible (Area Theorem) masses of the black holes.

QUANTUM GRAVITY

The unification of gravitation physics with quantum physics or the construction of a completely new theory incorporating both is one of the greatest challenges in theoretical physics. The challenge confronts us not so much because of the possibility of immediate experimental test (simple order-of-magnitude estimates indicate that laboratory tests of a quantum theory of gravity are not likely within the decade covered by this report); rather, the challenge of quantum gravity confronts us, first, because we observe a system for which we can be sure quantum gravity is important. This is the universe itself. Quantum gravitational effects are significant in the extreme conditions of the big bang, and there can be no understanding of the complete history of our universe without an understanding of quantum gravity. Second, the present vision of a unity of all particle interactions will not be complete until gravity is incorporated in that unity. Indeed, it may be that gravity enters in an essential way into any fundamental understanding of matter. Third, there are some explicitly observational problems that will require a deeper theory, as we shall see below.

There is no lack of issues in quantum gravity; throughout the field there are unresolved problems and issues of principle. Working out the quantum mechanics of Einstein's classical theory would seem a reasonable starting point in the study of the quantum theory of gravitation. Not only are we unable to calculate effectively with the resulting theory (it is not renormalizable), but fundamental issues such as identifying the variable that plays the role of time and the construction of the Hilbert space of states are still not satisfactorily resolved.

It may be that the Lagrangian for general relativity, so unique and successful in the classical regime, does not correctly describe the

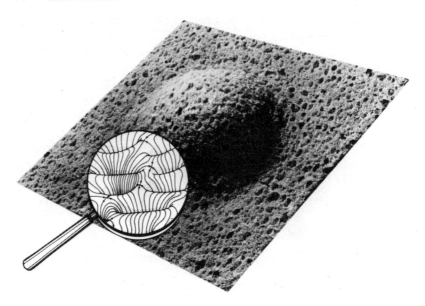

FIGURE 9.1 Space-time foam. On length scales of the order of $(\hbar G/c^3)^{1/2} \sim 10^{-33}$ cm space-time undergoes enormous fluctuations in curvature with associated energy density $c^5/\hbar G^2 = 5 \times 10^{93}$ g/cm^3. Of the same order of magnitude is the negative energy density due to gravitational attraction of the wormholes. Space-time foam illustrates the geometric approach to quantum gravity.

quantum mechanics of space-time on distances of 10^{-33} cm (see Figure 9.1); rather, it may be an effective model good only on longer scales. Perhaps the correct Lagrangian is one in which gravity is unified with matter theories, or perhaps there is no gravitational Lagrangian at all. Lagrangian theories of gravity tend to share common problems. Perhaps the most important is the problem of the cosmological constant, or energy density, of the vacuum state. Calculation of quantum corrections to typical field theories suggests a cosmological constant of the order of unity on the Planck scale; observation tells us it is 10^{120} times smaller. Understanding these 120 orders of magnitude is one of the most significant challenges confronting any quantum gravitational theory.

It may be that local Lagrangian field theory is not the correct approach to quantum gravity. Perhaps, as some believe, the basic quantum quantities are not the variables describing a space-time continuum but a more discrete structure. Finally, it may be that the laws of quantum mechanics themselves require modification in the extreme physical regime where quantum gravitational effects are important.

There are many avenues of approach that promise to shed light on a quantum theory of gravity and its applications. A partial list of them includes the canonical approach, covariant perturbation theory, Euclidean quantum gravity, quantum field theory in curved space-time, geometrical quantization, twistor theory, discrete gravity, curvature-squared theories, nonlinear quantum mechanics, spin networks, induced gravity, asymptotic quantization, quantum cosmology, supergravity theories, Kaluza-Klein theories, and superstring theories. One could perhaps even attempt to assess their prospects viewed from some present perspective. To do so, however, would not provide a guide for the future of the area. There are many diverse approaches because there are many ideas and deep unsolved problems. There is no obvious single approach, and there should be none at this stage. The best hope for substantial progress is to encourage a variety of approaches and to encourage cross-fertilization between them and with other relevant areas—quantum field theory, particle physics, and mathematics, on the one hand, and cosmology and astrophysics on the other. One can expect developments in the area to proceed by fits and starts. New ideas will be proposed, tested, and either abandoned or added as pieces of an as yet incomplete structure. New techniques will produce new objectives, and new objectives will produce new techniques. Taking greater risks will be necessary to support diversity and encourage innovation, but the payoff will be a deeper understanding of perhaps the most fundamental problem of physics.

ASTROPHYSICAL PROPERTIES OF NEUTRON STARS AND BLACK HOLES

Work should continue on modeling of astrophysical properties for neutron stars and black holes. Here the relativity physics is fairly well understood, but the interaction between general relativity and other phenomena such as hydrodynamics, electrodynamics, and radiative transfer remains to be understood in detail. The construction of models for active galactic nuclei and quasars, both of which involve accretion onto black holes, and their confrontation with observation, is an active and quite challenging problem in relativistic astrophysics. A crucial lack is the absence of currently available observational means to distinguish between black-hole models (see Figure 9.2) and other sorts of models.

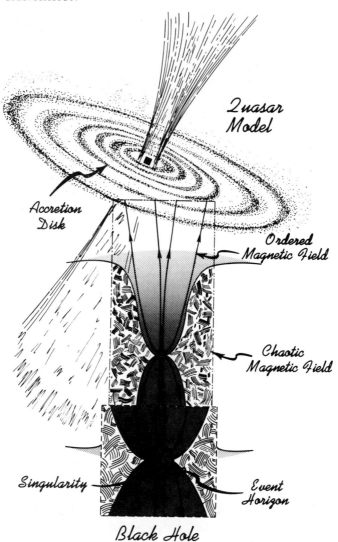

FIGURE 9.2 One possible model for generating the jets seen coming from some radio galaxies and quasars. An accretion disk orbiting a supermassive (10^9 M_{Sun}) black hole deposits chaotic magnetic field onto the hole, which "cleans" the magnetic field lines that thread it. The ordered field interacts with the hole's rotation-induced gravitomagnetic field to produce ~10^{20}-V potentials that accelerate relativistic particles out the poles, forming jets. This model exemplifies the complexity and variety of physics possible for black holes in an astrophysical setting and the importance of more detailed observations.

COMPUTATION

The Einstein equations form a difficult system of nonlinear partial differential equations. Lacking a general solution by analytic means, we must rely on numerical solutions for many applications of the Einstein equations, notably for gravitational collapse, black-hole collisions, and inhomogeneous cosmology. Great progress has been achieved in the last decade on numerical relativity using large-scale computers, but the equations are difficult enough that the significant computational problems remain untouched. The most difficult problems, those involving full general relativity in three space dimensions and one time dimension, will be in reach with supercomputers of the capability projected for the next decade, although substantial development of numerical algorithms will also be required. An example is the problem of the black-hole binary, in which one follows the orbital decay and final coalescence of two black holes in a binary system with energy loss by gravitational waves. As possibly the strongest gravity-wave source in the universe, this mechanism holds great promise for testing relativity in the regime of highly dynamical strong fields, if the wave forms can be detected and measured.

A second important use of computers in relativity is for symbolic manipulations. The analytic computations in relativity are often extraordinarily intricate, and computer assistance is often useful or even essential. Symbolic manipulation packages for algebra and calculus have gradually become more and more significant owing to the increased availability of hardware and to great advances in software algorithms for symbolic manipulations. The development of supercomputers, and provision of access to them by researchers, will play an increasingly important role for research on certain important problems in gravitation theory.

NEW KINDS OF EXPERIMENTAL TESTS

Solar-system tests of relativity are now approaching a precision of one part in 10^3 of the first post-Newtonian terms in effects such as time delay and light bending. To reach the level of second-order post-Newtonian effects will require a further factor of 10^3 improvement; as we have seen (see section on Measurement of Second-Order Solar-System Effects in Chapter 3), experiments at this level are under study. Further theoretical work on second-order post-Newtonian effects, in general relativity and especially in alternative theories, will be needed. New theoretical proposals may also be needed to interpret current tests of the R^{-2} law for Newtonian gravity in the laboratory and on Earth, over ranges of millimeters to kilometers. For instance, axion forces that arise in certain field theories of elementary-particle physics give some additional motivation for such experiments and suggest possible anomalous effects, such as spin-dependent forces and forces that violate time-reversal invariance.

Relativity predicts the evolution of the universe, and, therefore, observations in cosmology may someday be used to test the theory. At present, the theory is used to interpret the data rather than the data used to test the theory. However, as cosmological data become more extensive and precise, the situation could be reversed. Analysis of the consistency of cosmological models with observations, therefore, continues to be an important theoretical question.

Current speculations in quantum gravity suggest exotic effects, such as violation of CPT invariance, evolution of pure quantum-mechanical states to mixed states, and baryon decay mediated by gravitational effects. Supergravity yields a number of effects of its own. At present all these seem far too weak to measure, but the possibility exists that some such effect will turn up that is within experimental reach. So far, no actually or potentially observable phenomena in high-energy physics have been tied to gravity, but modern Grand Unified Theories are importantly influenced by virtual processes that transpire at the grand unification mass scale, which may be only 2 to 4 orders of magnitude below the Planck mass scale of quantum gravity. One may optimistically hope for direct connections between the observable phenomena of high-energy physics and quantum gravity sometime in the next decade or two.

COMMUNICATION WITH OTHER SUBFIELDS: GRAVITATION EXPERIMENT, ASTRONOMY AND ASTROPHYSICS, FIELD THEORY AND ELEMENTARY-PARTICLE PHYSICS, PURE MATHEMATICS

General relativity theory has experienced a period of great growth over the past 20 years. An important stimulus for this growth has been the interchange of ideas and problems with other subfields. The discovery of pulsars and quasars by astronomers has focused much attention on theoretical studies of neutron stars and black holes. In turn, the discovery and observation of gravitational-wave sources may provide a new window for astronomical observations of compact objects. Tests of relativity have stimulated much work on alternative theories as well as on the observable predictions of the Theory of

General Relativity, and tests have now ruled out important classes of alternative theories.

The example of general relativity has provided an important stimulus over the past 60 years to field theory; and in particular the supergravity theories and Kaluza-Klein theories, considered important hopes for unification, grew out of general relativity. The problems of frontier particle physics have to a significant extent become those of gravitation physics. In turn, developments in field theory have given rise to new directions in gravity by providing new techniques and new theories in which gravity plays a part. One can expect this close relationship between general relativity and particle physics to grow even more rapidly in the coming decade.

Communication with pure mathematicians led to the proof of the Positive Energy Theorem, one of the most important results in gravity theory in the past decade. Modern ideas from algebraic geometry have significantly influenced and contributed to the progress of the twistor program and to the study of complex spaces at asymptotic null infinity. One can also expect this close relationship with mathematics to grow as mathematical tools become even more important in the exploration of theoretical ideas.

Continued strong relations of gravitation theory with other subfields such as those just mentioned will be essential for its continued vitality, and indeed for the vitality of theoretical physics as a whole.

10

Recommendations

SPACE TECHNIQUES

- An important part of general relativity remains completely untested: the prediction of gravitomagnetic effects, though exceedingly small in the solar system, should be checked experimentally.

The National Aeronautics and Space Administration's (NASA's) Relativity Gyroscope Experiment (Gravity Probe B) is currently our best hope of detecting such an effect—the dragging of inertial frames by the rotating Earth. The experiment calls for a level of technical sophistication not yet achieved in a spaceborne instrument. We are pleased to note that NASA has initiated the first phase of a two-stage program designed to accomplish this mission.

- The highly successful use of solar-system ranging experiments to test general relativity should be continued.

Ranging to planetary landers and orbiters has been particularly fruitful, and no such opportunities should be missed. The Mars Observer mission appears to be the first such opportunity if an accurate dual-frequency ranging system is included. It also is of great importance to keep improving the solar-system model with laser ranging to the Moon and radar ranging to the planets. These techniques are extremely cost-effective means for increasing the stringency of solar-system tests of general relativity.

- Two frontiers in gravitation research are the detection of gravitational radiation and the testing of general relativity in second order. Promising ideas for space experiments in these areas should be encouraged and studied.

Current concepts that warrant further study as possible future NASA missions are a long-baseline ($\sim 10^6$-km) laser interferometer in solar orbit to detect gravitational waves in the important millihertz frequency range and a precision optical interferometer (POINTS) capable of testing relativistic light deflection by the Sun to second order. NASA currently has a proposal to send a precision clock into a near-Sun orbit (STARPROBE) to measure the gravitational redshift to second order, thus making a new (clock) measurement of the PPN parameter β.

GROUND-BASED TECHNIQUES

- The strain amplitude sensitivity of interferometric gravitational-wave detectors can be increased by 2 or more orders of magnitude by the construction of baselines with lengths of ~ 5 km. This facility offers the opportunity for a breakthrough in gravitational-wave detection and should be pursued vigorously.
- Bar detectors are today the most sensitive gravitational-wave detectors. A diverse research program should enjoy continued support, with due attention being given to critical technologies. However, systematic observations should play an increasing role as a guide in the development of bar detectors.
- Pulsar observations have provided an impressive demonstration of gravitational-wave damping in the binary pulsar and significant upper limits for microhertz gravitational waves in the millisecond pulsar. Searches for and observations of pulsars and other compact objects, especially in binary systems, should be given high priority.
- Laboratory experiments continue to play a role in gravitation research by testing with increasing precision the basic principles and predictions of gravitation theories. The fundamental nature of this work more than justifies its small cost.

GRAVITATION THEORY

- Continued support for theoretical research is crucial to the health of gravitation physics. The essential prerequisites for a strong theory program are (a) support for a diversity of high-quality research areas, (b) availability of means for communication among theorists and also with scientists in other specialties, and (c) adequate opportunity for entry into the field by talented young people.

- The strong relations of gravitation theory with other areas including particle theory, gravitation experiment, astrophysics, and pure mathematics are important to the field and should be fostered.
- Large-scale computation is playing an increasing role for certain problems in gravitation theory, as in many other fields. We welcome the initiatives currently under way to improve the access of physical scientists to supercomputers and smaller computers.

III

Cosmology

Cosmology, the study of the universe as a whole, provides the canvas on which the detailed nature of the physical world is painted by the other fields of physics. This canvas is the space-time framework upon which all our physical theories are constructed. The question of boundary conditions in both space and time (e.g., the issue of origin) is ultimately a cosmological one.

A second feature of cosmology that endows it with fundamental importance as a field of physics is the fact that the properties of matter are studied under the most extreme conditions, from the unimaginable densities and temperatures of the early universe to the near-perfect vacuum of intergalactic space. By comparison, experimenters in terrestrial laboratories can only test our physical theories over a narrow region of their supposed range of validity.

But this potential for expanding our understanding of physics comes at a price—the uncertainties introduced by the remoteness of our cosmological laboratories. Because only passive experiments (i.e., observations) are possible, theory must play a particularly critical role in the planning of experiments as well as the interpretation of data and the distillation of knowledge. An additional difficulty arises because of the uniqueness of the universe, which prevents us from determining whether our universe has a particular property by chance or by necessity. Related to this problem is our inability to isolate the system under study; indeed the observer is inseparable from, and a product of, the system and processes being investigated.

During the past two decades, cosmology has undergone a revolution because of our increasing ability to observe the universe as it is now and as it was in the remote past. We have extended the horizon of our knowledge back in time, through the era of the quasars to that at which the microwave background photons were released—a time when the density of the universe was 10^9 times higher than it is now. And relic nuclei allow us to see back even farther, to a time when the universe was only a few seconds old. Currently, theorists are attempting to study still earlier times, by applying new ideas from particle physics to the universe at age 10^{-35} s. A major objective for cosmology is to extend and broaden our physical understanding of the early universe.

Equally exciting is our rapidly growing knowledge of the local universe, out to say 10^8 light years. Major advances in astronomical instrumentation and data-processing techniques have led to more detailed studies of the physics of galaxies and clusters of galaxies—data vital to understanding the origin and evolution of these basic elements of our universe. Important puzzles, such as the nature of a probable dark-matter component and the physics of galactic nuclei, are stimulating a burst of theoretical and observational activity. We can expect this area, so rich in basic phenomena, to continue to grow and flourish, aided greatly by new layers of knowledge from major new astronomical instruments.

11

Introduction — The Standard Model

The recent renaissance in the development of cosmology has occurred mainly within the context of the hot big-bang models, whose governing gravitational equations were derived more than 50 years ago. These are the models currently employed by most cosmologists because they are the simplest and most natural ones in accord with the observations. For example, big-bang models are compatible with (1) the isotropy of radiation backgrounds and galaxy counts, (2) the galaxy redshift-distance relation, (3) the observed ages of the oldest stars and meteorites, (4) the cosmic microwave background radiation temperature of the universe (~ 3 K), (5) the present mean density of matter, (6) the rate of expansion and deceleration of the universe, and (7) the abundance of primordial elements. The renaissance in cosmology was sparked by the realization that the microwave background radiation and the light-element abundances are remnants of a hot big bang, but it has been driven by the successful application of a broad range of observations and theory to difficult cosmological problems. Currently, revolutionary ideas concerning the relationship between microscopic physics and the large-scale structure and evolution of the universe are being actively studied and tested.

Figure 11.1 shows the past history of the universe, according to a standard big-bang model and including recent ideas from particle physics. The contribution of the various particles to the mass-energy density is plotted versus time since the extrapolated epoch of infinite

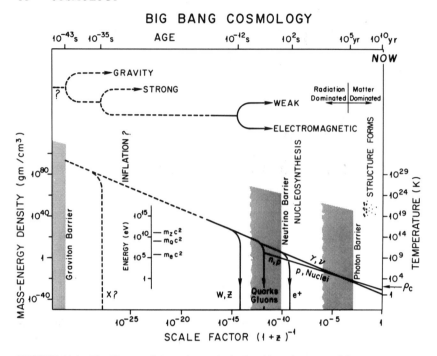

FIGURE 11.1 The history of the universe is depicted here in terms of the mass-energy density of the different types of particles that were present at various epochs. At early times, many types of elementary particles (X, W, Z, quarks, gluons, μ, τ, e, γ, ν, ...) existed. Most disappeared because of particle-antiparticle annihilation when the particles' kinetic energies became less than their rest-mass energy. Neutrons (n) and protons (p) were produced from quarks at about 10^{-5} second, and light nuclei were produced from nucleons at about 10^2 seconds. The three barriers indicate the epoch beyond which we cannot "see" via each of the three types of particles. In addition, at energies higher than that available at accelerators we have little direct knowledge of the laws of physics (hence, the dashed curves).

density. It is this mass-energy density that controls the expansion rate in the standard model, so the distance scale factor $(z + 1)^*$ can also be shown. Most of what we know about the universe comes from astronomical observations of optical and radio sources at the extreme right-hand edge of Figure 11.1, between the present and a scale factor of $z + 1 = 2$, corresponding to a time when the universe was about one half its present age.

---

*z = cosmological redshift = $[\lambda(\text{source}) - \lambda(\text{lab})]/\lambda(\text{lab})$. So, the distance scale, which is proportional to wavelength, changes as $z + 1$.

In the hot big-bang model the 3-K microwave background is a remnant of primordial blackbody radiation. However, the radiation is strongly scattered before the photon barrier at $z \sim 10^3$ because at earlier times the radiation temperature was high enough to ionize hydrogen and permit Thomson scattering. This strong coupling of radiation and matter before the photon barrier also tended to keep the matter from clumping into stars, galaxies, or larger systems bound by gravity. Decoupling of the matter and radiation allowed the process of galaxy formation to begin, leading eventually to the complex large-scale structure seen today. Since the 3-K photons were last scattered at the photon barrier (unless matter is reionized), their current properties carry the imprint of this epoch ($z \sim 10^3$, $T \sim 10^4$ K, age $= t \sim 10^5$ years).

Note the tremendous range of physical conditions that the model encompasses, with densities reaching 10^{94} g/cm^3 at the Planck era where the unknown laws of quantum gravity prevail. The bold extension of our present knowledge of physics into the early universe represents the greatest extrapolation in all of science. However, this extrapolation provides a unique opportunity to derive observable consequences from the laws of physics that we imagine to operate under such conditions. As we shall indicate below, relic particles (produced in the early universe and still present today) may provide the key, or perhaps the spectrum of residual density fluctuations will be our deepest probe. Even more likely, the observational breakthrough to the particle-physics era will come in some entirely unanticipated way; new ideas are currently appearing at a rapid rate.

12

Highlights

BIG-BANG NUCLEOSYNTHESIS

At present, the relics of the big bang that provide the most information about the early universe are certain light nuclei, such as deuterium (D) and ^{4}He. Calculations of their production in the early universe are based on measured nuclear cross sections and rely heavily on quantitative details of the cosmological model. A few minutes after the origin of the universe, conditions of temperature and density were appropriate for the fusion of protons and neutrons to form nuclei of light elements. Deuterons formed first, but the fusion reactions ran rapidly toward ^{4}He because of its much greater stability. The amount of ^{4}He produced depends essentially on two factors: the density of baryons (neutrons and protons) and the universal expansion rate at the epoch when the temperature dropped to $\sim 10^9$ K ($t \approx 3$ min). The baryon density at $T \sim 10^9$ K can be computed from the present baryon density and the present temperature of the background radiation; and the expansion rate can be calculated for isotropic, homogeneous cosmological models provided that the number of species of light particles is known. Hence precise predictions of the ^{4}He abundance can be made; the calculated value lies in the range of 23-27 percent by mass. This agrees with the solar value and with the abundance found on old stars and in the interstellar medium, after correcting for the ^{4}He made in stars. The fact that the predicted abundance of ^{4}He agrees with the

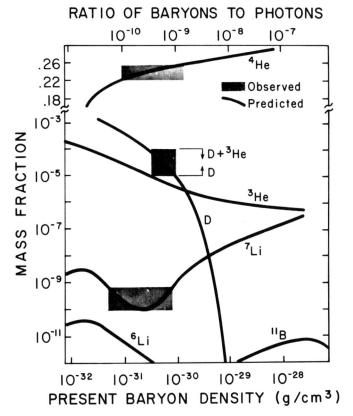

FIGURE 12.1 Isotopic abundances compared with predictions of the standard big-bang model (black curves). Shaded areas indicate observed abundances for ^{4}He, deuterium, ^{3}He, and ^{7}Li, which all show remarkable agreement with theory for a baryon-to-photon ratio in the range 10^{-10} to 10^{-9}.

observed value provides the best evidence for the validity of the standard big-bang model at these early times.

Even more has been learned from studies of light-element abundances. While the ^{4}He abundance is not a strong function of the density of baryons, the small residual D abundance depends sensitively on this quantity, being relatively larger for lower baryon density. Figure 12.1 shows the predicted primordial abundances of several light nuclei, as functions of the present baryon density—a poorly known cosmological parameter. The observed abundances are shown by shaded rectangles, with ^{7}Li being a recent addition. The agreement with predictions is striking and suggests a present baryon density of 3×10^{-31} g/cm^3. This density is 1 or 2 orders of magnitude less than the density required to close the universe, that is, to stop the current expansion and cause recollapse. It also may be less than the density required to explain the observed dynamics of large clusters of galaxies. Thus, suspicion is rising that the long-sought invisible (sometimes called missing or dark) matter is something other than baryons. We return to this point below in the section on Invisible Mass.

One might suppose that the observed light nuclei were produced by much later astrophysical processes, making the agreement in Figure 12.1 fortuitous. At this point we know that these light nuclei cannot be produced by the collisions of cosmic rays with the interstellar gas and that no other production mode has been found for D. Thus, the deuterium abundance is particularly important.

The sensitivity of the production of ^{4}He to the expansion rate of the universe at $t \sim 3$ min has allowed constraints to be placed on other physical parameters. For instance, if more than a few types of neutrinos exist, the expansion rate would have been greater, resulting in excessive production of helium. Also, if the gravitational "constant" had been different at that early epoch ($\dot{G}/G \neq 0$), the expansion rate and the helium production would have been altered. Finally, the universe could not have been very anisotropic at $t = 3$ min, because that would also have increased the average expansion rate.

LARGE-SCALE PROPERTIES OF THE UNIVERSE

The general expansion and deceleration rates of the universe have been a central focus of cosmology for the past 30 years. Recent work has narrowed the uncertainty in Hubble's constant, a measure of the current expansion rate ($H_0 = 50$ to 100 km/s per Mpc*), but the deceleration parameter q_0† remains poorly known. The classical methods to study the geometry of space-time use visible galaxies and radio sources as coordinate measures. Usually, source intensity is used as a measure of distance, but this requires a knowledge of time dependence of the source luminosity and spectrum. The effects of source evolution have not yet been sufficiently well understood to permit a geometrical

---

* 1 megaparsec (Mpc) $\sim 3 \times 10^6$ light-years, roughly 1/5 the spacing between large galaxies.

† In the simplest big-bang models (pressure = 0, cosmological constant = 0) $q_0 = \frac{1}{2}\rho_{measured}/\rho_{critical}$. For $q < 1/2$ the universe is open and expands forever; $q > 1/2$ means our universe is closed and will recollapse. Measurements of the density ρ and deceleration q_0 of the universe are of major importance to cosmology.

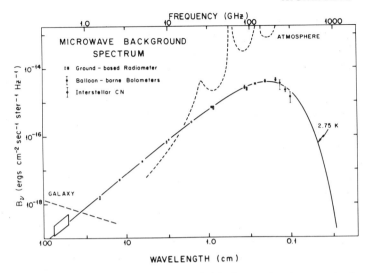

FIGURE 12.2 Measurements of the absolute flux in the cosmic microwave background radiation. Only the more accurate measurements at each wavelength are shown on this graph. The microwave photons excite rotational levels of interstellar CN, and the populations of these levels are measured by absorption of starlight. The weighted mean temperature from the results shown is 2.74 ± 0.03 K, but the error in the mean is questionable since systematic errors dominate statistical errors in these measurements.

measurement of the universal deceleration rate. Later, in Chapter 13 on Opportunities, we discuss briefly how improved detectors are rekindling interest in classical methods.

Our first direct evidence of large-scale behavior in the early universe came from measurements of the spectrum of the 3-K radiation. The hot big-bang model predicts a blackbody spectrum, with only small deviations. Despite repeated careful measurements, there is currently no evidence for significant deviations from a blackbody curve with a temperature of 2.75 K. Figure 12.2 shows the results, including recent ground-based data from an international collaboration and results of a balloonborne experiment using filtered cryogenic bolometers. Earlier balloon observations with a Fourier-transform spectrometer suggested spectral deviations near the blackbody peak; these are not confirmed by the recent data.

The distribution of extragalactic radio sources indicates that the universe is homogeneous and isotropic on large scales (distances of $\gtrsim 10^2$ Mpc). Measurements of the isotropy of the 3-K radiation confirm this to better than 0.01 percent. In the simple model, the 3-K photons were last scattered at $z \sim 10^3$ (the photon barrier), so the isotropy measurements argue that the universe was homogeneous and isotropic at that epoch. Einstein's major cosmological assumption of large-scale homogeneity and isotropy seems well justified.

Incidentally, there is an interesting noncosmological feature in the anisotropy of the 3-K radiation—the dipole effect. It arises from the Earth's motion through the radiation and measures our velocity relative to the reference frame of the radiation, assumed to be the same as that of matter at large distances. The inferred velocity of our galaxy is surprisingly large and suggests that we are being perturbed by local mass concentrations such as the Virgo cluster of galaxies and its surroundings.

Failure to observe a quadrupole anisotropy with an amplitude larger than 10^{-4} K provides an important constraint on homogeneous but anisotropic cosmological models. Such models are completely consistent with general relativity, and indeed the number of such models is much larger than the number of isotropic models. Nevertheless, observations of the isotropy of the background radiation, and the agreement of the predicted and measured abundances of light-element abundances, tightly restrict the range of possible anisotropic models.

Finally, the high degree of isotropy in the 3-K radiation raises a serious causality question. In the standard model, regions separated in the sky by more than ~ 1 degree were not yet causally connected at $z \sim 10^3$, the epoch of last scattering (assuming no reionization). How then did the photons coming from those regions manage to have the same temperature, to 1 part in 10^4? This long-standing problem with the 3-K radiation in the simple model may be solved by a fascinating new idea—the inflationary universe—discussed below in the section on The Inflationary Universe.

STRUCTURE IN THE UNIVERSE

The clumping of matter in the universe into galaxies, clusters of galaxies, and still larger structure is currently under intense scrutiny. Quantitative observational work has rapidly accelerated with new developments in detector and data-processing technology. Analyses of angular distributions of galaxies on photographic plates are now complemented by three-dimensional information from the first large-scale statistical samples of galaxy redshifts.* Redshift measurements require spectra, so they take much more time to obtain than do

---

* The redshift gives the recession velocity, which is related to the distance by Hubble's law $v = H_0 d$.

photographs, but redshift surveys yield a much clearer picture of the galaxy distribution and dynamics. On small scales (≤ 10 Mpc) the galaxy distribution approximates a scale-invariant fractal within which there is an occasional great cluster of galaxies. On larger scales one finds a complex pattern of superclusters, clouds, voids, and filaments of galaxies. There is considerable theoretical and observational activity devoted to tracing the evolution of this structure back to its origins. Evidence exists that radio sources, quasars, and perhaps also galaxies have changed appreciably between the epoch $z \sim 3$ and the present. But the burst of radiation that may accompany galaxy formation at an epoch somewhere between $z \sim 3$ and $z \sim 100$ has not yet been seen. Indeed, so little is known about the formation and development of structure in the universe that we are currently debating whether stars or large clusters of galaxies formed first.

Galaxies and clusters of galaxies may arise from small density fluctuations, $\Delta\rho/\rho$, in the early universe. Two limits can be set on the magnitude of fluctuations at the time of decoupling of matter and radiation. Since the current density contrast on the scale of clusters of galaxies is about unity, and gravity causes $\Delta\rho/\rho$ to grow as $(1 + z)^{-1}$, the perturbations at $z \sim 10^3$ should be $\Delta\rho/\rho \sim 10^{-3}$. Another limit comes from the search for small-scale anisotropy in the 3-K radiation, which is a probe of roughness on the $z \sim 10^3$ surface. At angular scales corresponding to the sizes of large clusters (a few arc minutes) no fluctuations are seen down to $\Delta T/T \sim 2 \times 10^{-5}$. Current results of isotropy measurements of the 3-K radiation are shown in Figure 12.3.

Under certain assumptions about the character of the fluctuations these two ways of estimating $\Delta\rho/\rho$ are in conflict. For example, adiabatic perturbations (favored by some models, especially those derived from particle-physics considerations) give $\Delta\rho/\rho \sim 3\Delta T/T$. Then the density contrast seen today ($\Delta\rho/\rho \sim 1$) implies $\Delta T/T \sim 3 \times 10^{-4}$ at $z \sim 10^3$. But the limits shown in Figure 12.3 at scales of a few arcminutes are ten times smaller. There are several ways out of this dilemma: make the perturbations isothermal, clump the matter by forces other than gravitational (e.g., by supernova explosions), or rescatter the 3-K photons from an intermediate screen of electrons at $z < 10^3$. A recent idea suggests that nonbaryonic, invisible matter (e.g., axions, photinos, or massive neutrinos) can become nonrelativistic and begin to clump before $z \sim 10^3$. Baryons then fall into these clumps after decoupling from the radiation. Thus, there is ample time for structure to form in the invisible matter, and baryonic matter and microwave photons can be very weakly perturbed at decoupling.

This is only one of the many ways that newly suggested particles

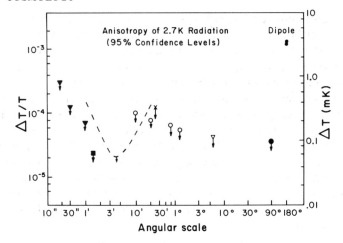

FIGURE 12.3 Current results of searches for anisotropy in the cosmic microwave background radiation. The only effect seen, so far, is the dipole, which is due mainly (and perhaps totally) to our velocity through the radiation. Various symbols denote different observational techniques. Generally, small balloonborne instruments are used at angular scales larger than 3 degrees, and ground-based radio telescopes are used at smaller angular scales.

have been used to try to solve certain cosmological problems. On the other hand, the universe is a good laboratory in which to try out the properties of new particles. For example, the various candidates for invisible matter have different clustering properties. Some can form seeds for structures in the matter; others can provide a smooth mass density to help close the universe. Thus, cosmological observations can place constraints on the properties and abundances of new kinds of particles.

INVISIBLE MASS

The dark-matter problem is not new to cosmology. Observations since the 1930s have indicated that the mass density of visible matter (stars and gas) is insufficient to close the universe or to explain the dynamics of large clusters of galaxies, and recently it has become apparent that the visible mass cannot account for the strength of the gravitational field in the outer parts of galaxies, as indicated by the motions of stars and by the concentration of plasma around some galaxies. The discrepancy between what is observed directly as visible mass and what is indicated by dynamical measurements ranges from a factor of 2 in our stellar neighborhood to a factor of about 5 in galaxies to a factor of 30 or more in clusters of galaxies. This interesting trend for the invisible-mass fraction to increase with scale is not understood.

With so much at stake, the search for the invisible mass is vigorous and extensive. Low-mass stars (mass $\sim 0.1\ M_{Sun}$) are now unlikely candidates because galactic halos do not exhibit excess brightness at $\lambda = 2\ \mu m$, where such stars are bright. Still lower mass objects ("Jupiters") are possible, since they are not luminous and hence are extremely hard to detect.

Black holes are popular candidates for the invisible mass; and again, they are hard to find, especially in isolation. Currently, we are not even agreed that a black hole has been identified, though there are several excellent candidates among the known x-ray sources. Massive black holes ($\sim 10^8\ M_{Sun}$) are suspected as the "central engines" in active galaxies and quasi-stellar sources. Also, primordial black holes with masses down to 10^{15} g could exist and easily have escaped detection. (Those with masses below $\sim 10^{15}$ g are predicted to have evaporated by now by the Hawking process; see section on Quantum Particle Creation by Black Holes in Chapter 8.) Theoretical studies have taught us much about the astrophysical and relativistic properties of black holes, but we still do not understand how (if at all) such objects act as the powerhouses for active galactic nuclei or whether it is reasonable to assume that black holes might have existed in great numbers in the early universe. Their contribution to the invisible mass remains unknown.

Since the dark-matter candidates mentioned so far are made from baryons, the nucleosynthesis constraint on baryon mass density has strong implications here. Stars, "Jupiters," and even black holes born after big-bang nucleosynthesis are included in the baryon density constraint noted in Figure 12.1—a density far short of that needed to close the universe. Thus, nucleosynthesis argues that one should look for nonbaryonic dark-mass candidates as a means to achieve closure density, $\rho_c \approx 10^{-29}$ g/cm^3.

Several, yet unobserved, elementary particles are being proposed as dark-matter candidates. Already before reports of measured neutrino mass came from the Soviet Union, cosmologists had speculated about massive neutrinos as a source of nonbaryonic mass density. If neutrinos have a rest mass of only a few electron volts, the thermal neutrinos produced in the hot big bang would dominate the mass density of the universe today. Although the reported measurement of neutrino mass is still controversial, it ushered in a flurry of theoretical activity resulting in even more invisible mass candidates. Axions were mentioned earlier as possible seeds for galaxies; they are light pseudoscalar particles produced during the transition from quarks to hadrons that preceded primordial nucleosynthesis. Other new candidates are suggested by supersymmetric particle theories, which give partners such as the photino and gravitino to currently known particles.

The virtually unconstrained richness of particle theory at very high energies can be expected to breed many invisible matter candidates. However, some constraints do exist. To help bind galaxies a relic particle must have sufficient abundance today and must have become nonrelativistic so that gravitational clumping could take place. Also, the mass of any fermion candidate must be greater than the phase-space limit provided by the exclusion principle.* Some proposed particles have natural clustering scales that can be compared to observed structure, but it is still controversial which clustering lengths give the best fit to the phenomena.

COSMOLOGY AND GRAND UNIFICATION

A recent dramatic development in theoretical physics was the realization that the early universe is a useful laboratory for the testing of particle physics; conversely, new ideas in particle physics can be applied to some fundamental cosmological questions. Most interest has focused on an epoch when temperatures were high enough ($T \gtrsim 10^{27}$ K or 10^{15} GeV) to possibly induce grand unification of three fundamental forces—the strong, the weak, and the electromagnetic. An early success of this idea was to provide a possible explanation of the puzzling asymmetry in the abundance of matter and antimatter in the universe, amounting to about one excess baryon (matter) per 10^9 photons. The standard cosmological model gives no clues, but Grand Unification Theories (GUTs) contain the necessary ingredients to answer this fundamental question. GUTs can be asymmetric with respect to particles and antiparticles, and they violate baryon conservation, producing a net baryon number in a universe that initially had equal numbers of baryons and antibaryons. The process occurs at a temperature corresponding to the rest-mass energy of the X boson (see Figure 11.1 in Chapter 11) which is responsible for the interconversion of quarks and leptons. Currently, particle experiments do not constrain the parameters of these theories nearly enough to allow an exact prediction of the baryon-to-photon ratio, although the detection of

---

*Roughly, $m^4 > \rho \hbar^3 v^{-3}$, or $m > 20$ eV for typical densities and velocities inside a galaxy.

proton decay would at least provide evidence that grand unification does occur at high energy. Thus, particle theory has provided a possible physical explanation for a fundamental cosmological property previously assigned to arbitrary initial conditions.

Another important consequence of GUTs is the possibility of producing magnetic monopoles from singularities in the scalar (Higgs) fields invoked to generate particle masses. This could have occurred at the GUT era in the early universe, as regions with arbitrary alignments of the Higgs fields came into causal contact. In fact, in the simplest big-bang models far too many monopoles would have been produced; their present mass density would dominate the universe and cause excessive deceleration of its general expansion. This problem may be solved by a revolutionary idea that introduces into the early universe a process called inflation.

THE INFLATIONARY UNIVERSE

An ingenious way has been found to avoid the problem of excess magnetic monopoles emerging from the GUT era and to explain some older cosmological puzzles as well. The idea is that if scalar fields exist, their vacuum expectation value could provide a contribution to the mass density that remains constant in time, like the effect of the cosmological constant first introduced by Einstein. During the time following the GUT era when vacuum expectation energy dominates, the universe expands much faster than in the usual big-bang models, and this exponential expansion drastically dilutes the density of monopoles. The inflationary epoch must terminate, at least by the time of primordial nucleosynthesis, so that big-bang cosmology reigns during its successful epochs. According to current ideas, inflation ends when a lower (zero) energy state becomes accessible to the scalar fields as the universe cools by expansion.

As noted earlier a particularly vexing problem with simple big-bang models is that regions of the universe having the same properties (radiation temperature, for instance) have never been in causal contact at the time we observe them. To explain the observed uniformity, one can invoke special initial conditions or quantum processes in the mysterious Planck era, but the inflationary model provides a specific alternative mechanism. In this picture, our entire observable universe is embedded in a larger region that grew from a single causally connected piece during the era of exponential expansion.

Inflation also provides a possible way of understanding the flatness question, which basically asks: Why is the universe so close to a balance between its kinetic energy of expansion and its gravitational binding energy? Considering the huge range of densities encompassed by the expansion, exceedingly fine tuning of this energy balance was required to allow the universe to reach its current state. The inflation picture explains this, again because of the enormous expansion factor. Indeed the model predicts a flat universe, which (for zero cosmological constant) has the current mass-energy density in the universe exactly equal to ρ_c, the critical closure density. At present, the study of the inflationary class of big-bang models is one of the most exciting areas within the rapidly growing union of theoretical particle physics and cosmology. But experimental support is needed; the discovery of the Higgs particles that produce the vacuum energy, for instance, would place inflation on much firmer ground.

GRAVITATIONAL LENSES

In 1979, an example of the long-predicted gravitational lensing was discovered. Multiple images of a quasar were formed by the bending of light in the gravitational field of an intervening group of galaxies. Such alignments are not so rare as one might think, because of the extreme distances to the quasars; six examples have been found to date. Cosmologists are intrigued because detailed geometric-optics calculations of the paths have led to the possibility that the distribution of mass within the lensing system can be studied. In addition, if the quasar's luminosity varies with time, the different delays along the paths to the different images provide an additional scale, which in principle allows a determination of the distance to the intervening galaxies and thus the Hubble constant, H_0 (see section above on Large-Scale Properties of the Universe). However, it may prove difficult to determine the properties of the lensing system well enough to realize this additional payoff.

13

Opportunities

OBSERVATIONS FROM SPACE

Cosmologists are eagerly looking forward to the observations from Earth orbit planned for the coming decade. A broad range of the electromagnetic spectrum will be covered by the proposed missions—the Gamma Ray Observatory (GRO), the Advanced X-Ray Astrophysics Facility (AXAF), the Hubble Space Telescope (HST), the Space Infrared Telescope Facility (SIRTF), the Cosmic Background Explorer (COBE), the Large Deployable Reflector (LDR), and an antenna in space to extend the Very-Long-Baseline Array. Previous astronomical satellites have brilliantly demonstrated that deeper exploration of space, in many spectral regions, holds great potential for making new discoveries, solving old problems, and raising important new questions. The recent results from the Infrared Astronomy Satellite (IRAS) are an example of the scientific power of well-planned observations from space. A list of planned studies and possible discoveries is long and exciting; we can mention here only a few examples directly relevant to current cosmological problems.

The many discoveries of IRAS highlight the untapped richness of the infrared sky, so long obscured by atmospheric absorption and emission. For cosmology the infrared region holds special promise because, as illustrated in Figure 13.1, this is where one may at last see the birth of galaxies. The burst of starlight expected to accompany galaxy

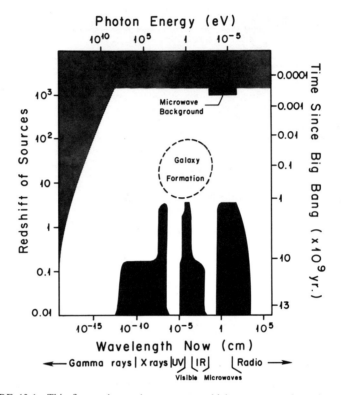

FIGURE 13.1 This figure shows the extent to which we can explore the universe throughout the spectrum of electromagnetic radiation (in terms of either the redshift of sources or, equivalently, how far back in time we see them). The darkly shaded areas show the extent of our present knowledge. The lightly shaded area shows the region that we can never view directly because the photons are either scattered by electrons or collide with other photons, producing electron-positron pairs. The dashed boundary surrounds the region where we may see galaxies in their early phase of development.

formation may have been redshifted by cosmological expansion into the infrared region. The detection and study of primeval galaxies will give us a major foothold in the little understood epoch between $z = 10^3$ and $z = 1$, and perhaps will also be the evolution of large-scale structure will be clarified. The National Aeronautics and Space Administration (NASA) is planning to capitalize on the success of IRAS by orbiting a cryogenic telescope with an aperture of about 1 meter—SIRTF. With pointing and imaging capability, SIRTF will be more sensitive than IRAS by factors of 10^2 to 10^3. Additionally, it will have more wavelength coverage and spectroscopic capability. A major part of SIRTF's scientific program will be a deep search for primeval galaxies.

Also of great cosmological significance is the planned role of the HST in the measurement of the extragalactic distance scale, the expansion rate of the present universe (H_0), and the deceleration parameter (q_0). High spatial resolution (better than 0.1 arcsec) and broad spectral coverage will allow more detailed observations of nearby and distant galaxies, leading to better understanding of the physical properties of galaxies including evolution. Thus, the HST is expected to play an important part in improving the classical cosmological observations over the next decade.

The HST's unique angular resolution and ability to measure redshifts at $z > 1$ suggest other observations of cosmological interest. A simple, but important, observation will be to see whether the shape of galaxies is evolving. Do the thin disks so prominent in most nearby giant galaxies persist back to $z \gtrsim 1$? Also, the HST will be a great help in charting the way for deep redshift surveys. Because of the large observing time required, the bulk of this work will be done from the ground (see the following section), but calibrations and minisurveys from space will be important benchmarks for these surveys. The HST will be our best means for studying supernova events in deep space. Currently, supernovae are being studied as possible cosmic distance indicators and as a possible alternative to galaxies as probes for measuring q_0. The advantage of using supernovae is that there is a good chance for theoretical understanding of the spectral and time dependence of their flux without needing to assume that they are standards of luminosity. One more example: HST's spectrometers operating at ultraviolet wavelengths (inaccessible from the ground) will be able to probe the thermal history of the intergalactic medium, which has been strongly influenced by the formation of structure in the universe. Thus, constraints can be set on the epoch of galaxy formation and on the nature of dark matter.

The COBE was designed specifically as a cosmological satellite, to make detailed measurements of the 3-K radiation and to look for an infrared background flux. High spectral accuracy will permit a search for distortions in the sensitive region over and around the blackbody peak ($\lambda \sim 2$ mm), and large-scale ($>7°$) anisotropy will be accurately measured at $\lambda = 9$, 6, and 3 mm. Because of limitations on the size of its antennas, COBE will not look for anisotropy at small angular scales.

AXAF was highly recommended by the report of the Astronomy Survey Committee as an instrument sure to make important contributions to broad areas of astronomy, including cosmology. Since the hot plasmas at the cores of some clusters of galaxies are strong x-ray emitters, AXAF will be able to make detailed measurements of these sources at redshifts of $z = 1$ to 2. Hundreds of sources per square degree are expected, with the number depending on the cosmological parameter q_0 and possible evolution. AXAF's ability to carry out detailed studies of these distant sources promises important new data from a little-known cosmological epoch. In addition, AXAF will provide much better data on nearby clusters of galaxies than was possible with the Einstein satellite. It will measure accurate temperature gradients as well as density gradients in the hot plasma in rich clusters. These will yield model-independent measurements of the gravitational potentials of the clusters and thereby trace the possible dark matter in the outer regions of the clusters.

The LDR is currently envisioned as a 30-m telescope, with diffraction limited at $\lambda > 30$ μm. Two important cosmological observations are being anticipated: a sensitive measurement of small-scale anisotropy in the 3-K radiation and a search for primeval galaxies at $z \sim 3$ using the reflector as a light bucket at $\lambda \sim$ 1-4 μm. Currently, LDR offers our best hope of pushing small-scale anisotropy measurements to levels of $\Delta T/T \leq 10^{-6}$, in pursuit of the primordial density fluctuation spectrum. Above the atmosphere LDR offers the low-noise, broadband capability needed for sensitive measurements near the peak of the spectrum at $\lambda \sim 2$ mm.

CONTINUED GROUND-BASED OBSERVATIONS

Most of what we know about the universe has been learned from interpreting observations made with ground-based instruments. Many of the data come from large telescopes at major observatories, but some important contributions have been made with small, special-purpose instruments. Always, the role of the theorist with a good understanding of the observations is an essential one, perhaps more so than in most areas of physics. The prospects for exciting ground-based work over the next decade are excellent; there is no shortage of important problems.

Because of the crowded schedule of broad-based science for the HST, only critical cosmological observations of relatively short duration can be made. Ground-based observatories will continue to be our main sources of data about the universe. The rapid pace of developments in extragalactic astronomy indicates that we are only just entering the age of discovery. The Astronomy Survey Committee discusses a broad range of opportunities for ground-based telescopes; here we emphasize only a few of particular current interest to cosmic physics.

Two major themes of current work are to measure q_0 and to understand the origin and evolution of large-scale structure in the universe. Recent redshift surveys of large numbers of nearby galaxies have greatly increased our understanding of kinematics and galaxy clustering in the local universe. It is important to extend this understanding to redshifts of $z \geq 1$ if possible. The joint distribution of galaxy redshifts and magnitudes will measure large-scale clustering of galaxies and afford a much clearer understanding of the evolution of structure in the universe, which depends on q_0 and on the nature of dark matter. Such a survey is technically feasible with current and planned telescopes. Curiously, the interpretation of data from a deep survey program would be limited in part by our lack of systematic, baseline knowledge of nearby galaxies. Such fundamental studies are well within the reach of present technology, but they have not been done. There is a perception among observers that such long-term programs, however important, cannot be undertaken because of uncertainties in funding and the allocation of telescope time.

Currently, our deepest look into the big bang is provided by measurements of the abundance of light nuclei. Astronomical observations of these abundances need to be extended to more sources and to even better accuracy. More theoretical work must be done to find and understand all possible astrophysical production and destruction mechanisms. As a cornerstone of our current hot big-bang cosmological model the nucleosynthesis argument must be as sound as possible.

Similarly, there is still much to be learned from further studies of the 3-K radiation. The spectrum near the blackbody peak needs to be measured still more accurately, large-scale anisotropy measurements can be improved (especially at millimeter wavelengths), and better polarization searches can be made. Fine-scale anisotropy measurements, of great importance to the understanding of primordial fluctuations, should be pursued from aircraft or balloons if necessary. Little is known about anisotropy on intermediate scales ($\sim 1°$); all angular scales are potentially interesting and should be probed to the highest possible precision.

The critical question of the nature of the dark matter that appears to dominate the present universe must be addressed by predicting and searching for signatures of the various candidates. Some possible signatures that have been suggested include x rays from accreting black holes, infrared radiation from very-low-mass stars, ultraviolet photons from the decay of massive neutrinos, direct detection of magnetic monopoles, and the photons from axion decay induced by magnetic fields. Theoretical studies will continue to impose constraints, such as the limits on cosmological monopole flux imposed by the existence of a galactic magnetic field (the Parker limit). Particle accelerators are not generally regarded as astrophysical observatories, but the discovery of a stable weakly interacting, massive particle could have a profound effect on cosmology.

PARTICLE PHYSICS AND COSMOLOGY

Conventional cosmology, if correct, places some important constraints on particle physics; examples are the allowed number of neutrino types and the allowed ranges of masses and half-lives of neutrinos. The new particle physics has generated some exceedingly stimulating ideas in cosmology and has great potential for influencing future thinking and directions; for example, the discovery of Higgs particles would be of major importance in lending credence to the inflation scenario. Within the decade the width of the neutral intermediate-vector boson Z^0 and the partial width due to neutrino pairs may be measured. Since the number of neutrino types affects nucleosynthesis, the measurement directly tests the big-bang model.

Many particle-physics experiments of interest to cosmologists do not use accelerators. One class of such experiments tests the predictions of theories, such as Grand Unification, which have implications in cosmology. Examples are the searches for proton decay and for an electric dipole moment of the neutron. Other experiments, such as those attempting direct detection of dark-matter candidates, offer the hope of a decisive resolution of important cosmological problems.

THEORY

Given the limited and indirect observational basis of cosmology, it is essential that theorists range broadly in their search for interpretations and for crucial observational and experimental tests. Fortunately, the field is sufficiently exciting to attract excellent theorists in graduate school and from other areas of physics and astronomy. It is impossible to anticipate where theory might go in the near future, but we briefly mention a few of the current promising ideas.

On the particle-physics side, the successful quantization of gravity seems essential for penetrating the mysterious Planck era. Perhaps only then will physics be able to address the question of initial conditions. Currently quantum gravity enjoys great popularity among gravitation, particle, and cosmological theorists. There has recently been much study of universes with more than four dimensions, motivated in part by supergravity theories, which attempt to unify gravity with the other three forces. (See the discussion under Quantum Gravity in Chapter 8.) An intriguing possibility is that these theories might lead to an understanding of the origin of space-time itself. Another difficult task is to develop the theory and consequences of symmetry-breaking transitions in the early universe. For example, the time-dependent transition that may cause inflation needs to be better understood.

On the astrophysical side, one attempts to understand the structure of the universe as it is now and to infer from that what it must have been like in the past. Essential to such a program are detailed studies of the complicated processes occurring during the nonlinear development of a multicomponent system of radiation and one or more dark-matter candidates. The processes must be understood from the present back to a time before the radiation decoupled from the matter. Such studies may invoke a wide variety of possible scenarios, but they must mesh with a rich texture of observations. In many cases extensive numerical computation is essential, and here a barrier to progress is the somewhat irregular and informal coupling of theorists to the frontiers of progress in computing technology. We also expect analytic methods to continue to provide new ideas and important guidance for observers.

Finally, we must bear in mind that the search for viable alternative cosmological models should continue. As an example, cold big-bang models in which the microwave background was produced by stars and thermalized by dust at an early epoch cannot be dismissed; they can give a present ratio of photons to baryons in agreement with the observed value. A major difficulty with such models is that no natural way to produce the observed deuterium has yet been found. Another class of nonstandard models are those that were initially chaotic rather than smooth. Is it possible that some process like particle production smoothed them out? What fraction of such models could evolve to resemble the present universe? What is the effect of an inflationary epoch on such models?

14

Recommendations

SPACE PROGRAM

• Cosmology is currently a data-starved science. We need to know much more about the universe now and at early times. To this end it is vital to maintain a vigorous program of space observations, such as that now planned by the National Aeronautics and Space Administration (NASA). The Hubble Space Telescope, the Cosmic Background Explorer, and the Gamma Ray Observatory are current missions of great interest to cosmology. Looking ahead, both the Advanced X-Ray Astrophysics Facility and the Space Infrared Telescope Facility will probe much deeper into the universe in their respective wavelength bands; important cosmological discoveries are quite likely from these instruments. Further off, the Large Deployable Reflector may be able to map the all-important small-scale anisotropy in the 3-K radiation, and a space arm for the Very-Long-Baseline Array will provide a fascinating look at details in the cores of radio galaxies.

• Scientific planning and instrumentation development for major space missions are often based on experiments carried out in balloons and aircraft, largely supported by NASA's suborbital program. The relatively low cost and quick turnaround time of these experiments permits diverse, exploratory research programs and realistic tests of developing instrumentation, especially new detectors. We urge NASA to consider some enhancement of this productive, cost-effective program and to continue its support of ground-based studies in support of space missions.

GROUND-BASED PROGRAM

- The revolution in cosmology over the past two decades has its roots in ground-based astronomy. Because of their intrinsic angular resolution and sensitivity to weak sources, large astronomical instruments such as the Very-Long-Baseline Array and the National New Technology Telescope (recommended by the Astronomy Survey Committee) are of central importance to cosmology; we strongly support these initiatives.
- The very productive U.S. program in astronomy is producing much of the basic data and many of the ideas underlying our current cosmological picture. It is essential that support of effective instruments and research programs be, at least, maintained as new initiatives are implemented. A strong scientific case can be made for increasing the level of support for U.S. astronomy and astrophysics.
- Several important problems in cosmology require systematic surveys of the properties and distributions of galaxies. These are expensive, long-term projects, perhaps best planned and managed by teams of scientists. We encourage the National Science Foundation (NSF) to consider how such projects might be organized and supported.
- We wish to note that the principal recommendation of the Elementary-Particle Physics Panel, a large new accelerator (the Superconducting Super Collider), has possible cosmological implications. The understanding of particle physics at the highest possible energies is necessary in charting the behavior of the early universe.

HUMAN AND COMPUTATIONAL RESOURCES

- Cosmology is currently done by a diverse group of scientists including astrophysicists, astronomers, relativists, particle physicists, nuclear physicists, and plasma physicists. This diversity is good for cosmology, which must draw from many fields of physics. However, as interest in the field intensifies, and more cosmology-oriented research groups form, the need for coordinated funding is becoming apparent. We encourage the NSF to consider how it might help to solve this growing problem.

- Many problems of great interest to cosmology require sophisticated computer technology; we think of N-body calculations, where N is large, of nonlinear hydrodynamic calculations, and of efforts to combine the two. We heartily endorse the NSF's recent initiative to help university-based groups to gain access to large computational facilities.

IV

Cosmic Rays

Cosmic rays provide our only direct sample of material from outside the solar system. Their composition reflects the nature of the nucleosynthetic processes by which all the elements of the periodic table are being constructed in the galaxy. In addition the cosmic rays are accelerated to relativistic speeds by processes in which nature concentrates vast amounts of energy in relatively few particles. These acceleration processes apparently take place on a wide variety of scales in astrophysical plasmas. Because some cosmic rays have energies higher than man-made beams of particles, they are also of interest for studying interactions of protons and atomic nuclei at ultrahigh energy.

Cosmic-ray physics is thus in essence an interdisciplinary field, touching astronomy and high-energy astrophysics, nuclear physics, plasma physics, and elementary-particle physics. It began as the study of energetic particles in the atmosphere, which we now know to be the products of nuclear interactions between the primary cosmic rays and air nuclei. In the past 35 years high-altitude balloons and spacecraft have carried instruments above most of the atmosphere, and the focus of cosmic-ray studies has shifted to the composition and energy spectra of the primary particles themselves, which includes atomic nuclei and electrons. The highest-energy cosmic rays, however, are still accessible only to surface experiments that can overcome the exceedingly low rate of these cosmic rays by exposing detectors of large area for long times. In addition, secondary neutrinos and muons are of great current interest for deep underground experiments, and there is an intense search for magnetic monopoles in the cosmic rays.

A major opportunity of the present decade is the ability provided by the Space Shuttle to place large detectors in space and to visit them subsequently for repair. By the early 1990s this capability will be supplemented by the Space Station, which will provide a permanent manned presence in space and permit routine maintenance and modification of orbiting instruments as well as assembly of instruments that otherwise would be too large to lift into orbit. The combination of Shuttle and Station will permit us to place new kinds of instruments in space, leading to new levels of precision of cosmic-ray instruments and extension of direct observation of the major cosmic-ray components by several orders of magnitude in energy. Ground-based detectors will remain the only source of information in the highest-energy regime, where galactic acceleration and confinement mechanisms probably fail, and one expects a transition to particles from outside our own galaxy. In both space and ground-based observations, instruments are now possible that will be capable of addressing some of the key astrophysical questions of processes of nucleosynthesis and particle acceleration, as well as questions of the physics of particle interactions at extremely high energies.

15

Overview

Because cosmic rays give us a direct sample of matter from some of the most energetic processes in nature and from distant regions of space, interest in the field remains high despite the difficulty of associating the particles with individual sources. Indeed, unraveling the physics of the acceleration of cosmic rays and of their propagation in the turbulent interstellar medium in order to discover the nature of the sources is a principal activity of the field.

The material of the solar system represents the local interstellar material as it was 4.6 billion years ago. The much younger cosmic-ray material, accelerated about 10 million years ago, provides a different sample of matter. In fact, recent observations suggest that cosmic rays may actually represent a more typical sample of the average interstellar medium than the solar-system material, which may have been contaminated by a nearby supernova explosion. The differences between cosmic rays and solar-system material are both significant and subtle. Understanding them will require more and better experimental data, perhaps new scenarios of nucleosynthetic processes, and a better understanding of the acceleration processes for the cosmic rays. In this way measurements of isotopic and elemental abundances will make important contributions to studies of the origin of the elements, a field that has only a limited amount of real data with which to check its theories.

There is much still to be learned about the nature of the material we observe at Earth as cosmic rays. Some elements (Ne, Mg, and Si) have been observed to have unexpected isotopic composition; models of cosmic-ray origin that could explain these compositions have been proposed, and observations of the isotopic composition of other elements are required to distinguish among these models. Studies of the abundances of the heaviest elements—platinum, lead, thorium, and uranium—are still primitive; much better observations are required if we are to determine the site and time scale of cosmic-ray nucleosynthesis. Observations of electron and positron spectra at higher energies and with greater precision are required if we are to determine the distribution of cosmic-ray acceleration sites in the local parts of the galaxy. Recent measurements of antiprotons at least require significant modification of simple models of cosmic-ray confinement in the galaxy and could also indicate more exotic sources; extension of antiproton observations to higher energies are required to distinguish among these possibilities. No antinuclei heavier than antiprotons have yet been observed in the cosmic rays, but if these searches could be extended at least two orders of magnitude in sensitivity, there is reason to believe that they would begin to be sensitive to extragalactic matter where these searches would take on much greater significance.

Nature demonstrates in many places its ability to accelerate particles. Solar energetic particles are accelerated at the Sun, particles are accelerated by the magnetospheres of the Earth and Jupiter, and under certain conditions particles are also accelerated in the interplanetary medium. The scale for acceleration of galactic cosmic rays is much larger, and far greater amounts of energy are involved. We see evidence for particle acceleration on an even larger and more energetic scale when we look at quasars and radio galaxies. Recently the binary object Cygnus X-3 has been observed with its characteristic 4.8-hour period by ground-based air-shower arrays in 10^{15}-eV gamma rays. If, as is likely, these are secondaries of nuclear collisions, this is good evidence of an energetic, distant source of cosmic rays in our galaxy. This could also imply a source of detectable neutrinos. Particle acceleration is evidently a common occurrence in a wide variety of astrophysical settings.

The total energy required to keep the galaxy filled with cosmic rays is enormous; it requires a substantial fraction of the energy released by massive stars such as supernova exploding at the rate of one every 30 years somewhere in the galaxy. The energy given to each cosmic-ray particle is also enormous; cosmic rays are truly exceptional—only one particle in 10^{11} in our galaxy becomes a cosmic ray, the most common cosmic rays have 10^{10} times as much energy as the thermal energy of a typical atom on Earth, and the most energetic cosmic rays are 10^{11} times still more energetic. Thus, cosmic rays play a crucial role in the energy balance of the interstellar medium.

Recent theoretical developments involving acceleration in various kinds of astrophysical shocks begin to make possible an understanding of the acceleration processes and, for the first time, lead to predictions. Measurements over the next decade should be able to test these theories through improved observations of the cosmic-ray energy spectra.

In their 10-million-year lifetime, the bulk of the cosmic-ray particles spiral around magnetic field lines, diffuse through the galaxy, and experience both nuclear and electromagnetic forces within a confinement volume whose size is still uncertain. Experimental data now put significant constraints on the details of the propagation and the conditions in the confinement region. The cosmic rays themselves also affect conditions in their confinement volume by ionizing material in molecular clouds, "blowing out" magnetic field lines, and generating secondary particles and photons through several different nuclear and electromagnetic processes. Major components of the diffuse radio and gamma-ray backgrounds are produced by cosmic rays. It is this intimate relation between the cosmic particle radiation and a broad range of physical processes that makes cosmic-ray studies such an important astrophysical discipline.

At some energy around 10^{14}-10^{15} eV or above, galactic acceleration and containment mechanisms must begin to fail. Nevertheless, the measured spectrum of cosmic rays extends to around 10^{20} eV without any sign of a termination. (See Figure 15.1.) Anisotropy of the cosmic rays increases continuously from a few tenths of a percent in amplitude around 10^{15} eV to more than 10 percent around 10^{19} eV. One recent analysis suggests that this is consistent with the increased difficulty of containing galactic cosmic rays and that extragalactic cosmic rays predominate only above 10^{19} eV. At the highest observed energies (about 10^{20} eV) it appears that cosmic-ray protons would be too energetic to be trapped in the known magnetic field of our galaxy or to survive energy loss by photoproduction on the relic blackbody radiation in propagation over cosmological distances. Cosmic rays of such high energies might come to us from our own local supercluster of galaxies, or they might come from the core of our own galaxy, bent back to the galactic plane by the (unknown) magnetic fields in a galactic halo. In any case these ultrahigh-energy cosmic rays are uniquely interesting and significant probes of cosmology and astrophysics.

The field of cosmic rays above 10^{15} eV forms a bridge between

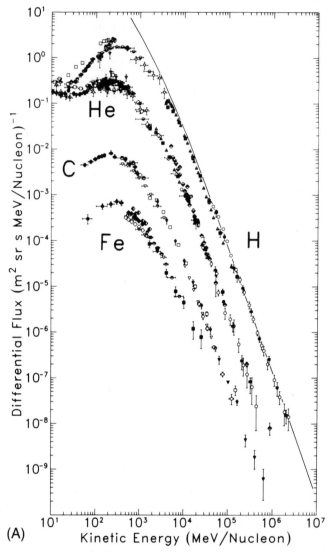

FIGURE 15.1 (A) The energy spectra of the cosmic rays measured at Earth. Differential energy spectra for the elements (from top) hydrogen, helium, carbon, and iron. The solid curve shows the hydrogen spectrum extrapolated to interstellar space by unfolding the effects of solar modulation. The turn-up of the helium flux below ~60 MeV n^{-1} is due to the additional flux of the anomalous ^{4}He component. From J. A. Simpson, Annu. Rev. Nucl. Particle Sci. *33*, 323 (1983). (B) The high-energy portion of the cosmic-ray spectrum (integral). Above 10^3 eV the composition is not yet well determined. The vertical bars indicate equivalent laboratory energies of existing and proposed (SSC) colliding-beam facilities.

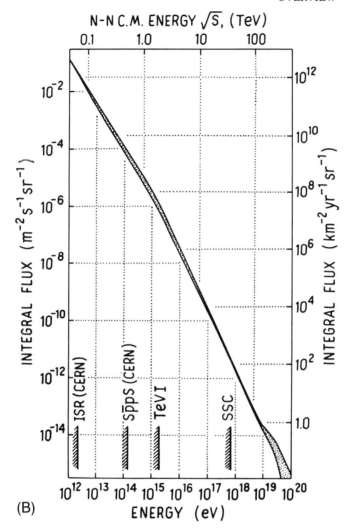

high-energy particle physics and experimental astrophysics. At and above these energies (the highest reached by the present generation of hadron colliding beams), the energy spectrum and chemical composition are accessible only by observations of cascades in the atmosphere with ground-based detectors.

Because the flux of the primary cosmic rays is so low at these energies, the relatively small detectors in spacecraft or balloons cannot intercept a large enough number for study. Large detectors can be exposed for periods of years on the ground to overcome this problem, but then the primary cosmic rays can only be seen indirectly through the shield of the atmosphere, which is some 10-15 interaction lengths thick. Ground-based detectors observe extensive air showers—the cascades of particles created by interactions of the primary cosmic rays high in the atmosphere. Because the energy is so high, the nature of strong interactions at higher energies (which determines how the cascades develop) must be inferred from extrapolations from accelerator data and from the indirect cosmic-ray data themselves. Because the interpretation of cosmic-ray cascades in terms of particle physics depends on the identity of the initiating cosmic ray (e.g., proton, carbon, or iron nucleus) and vice versa, our understanding of both areas is interrelated, and progress is made in an iterative, bootstrap manner as we move to higher energies. With the prospect of longer exposures in space we can expect the boundary between direct and indirect measurements to approach 10^{15} eV, and this will help to clarify the interpretation of the ground-based cascade studies at higher energies as well.

16

Highlights

This chapter discusses the problems being addressed by current and future cosmic-ray measurements. The general themes are organized by the history of cosmic-ray matter, starting with its synthesis, proceeding to its acceleration and propagation through interstellar space, and concluding with its interaction with matter.

To set the stage, we start with a list of the major discoveries of the last decade.

- New detectors have unambiguously resolved individual isotopes of neon, magnesium, and silicon. The resulting abundances show distinct quantitative differences from those found in the condensed bodies of the solar system, demonstrating conclusively that galactic cosmic rays are a sample of matter with a nucleosynthetic history that is different from that of the Sun. At the same time measurements of solar cosmic rays have provided some of the best measurements of the isotopic composition of the solar corona.
- Cosmic-ray abundances of individual elements heavier than iron have been successfully measured, despite the extreme rarity of these nuclei. The results indicate that the cosmic rays are not dominated by material recently synthesized in supernova explosions, as data suggested a decade ago, but may well be accelerated interstellar material, a conclusion that is consistent with the isotope measurements of the lighter elements.

- Measurements of the isotopes of the secondary element beryllium, in particular the abundance ratio of stable ^{9}Be to radioactive ^{10}Be, demonstrated that the cosmic rays that we observe today were accelerated on average 10 million to 20 million years ago and have propagated through interstellar material of mean density lower than the mean density of the galactic disk.
- The radial gradient of cosmic rays in the ecliptic plane of the heliosphere has now been measured. The gradient is less steep than some earlier models had predicted, and the edge of the modulation region [which had earlier been predicted to lie as near as 5 astronomical units (AU)] has been shown to be beyond 30 AU.
- A low-energy [tens of millions of electron volts/atomic mass unit (MeV/amu)] component with highly unusual composition was discovered. This anomalous component is rich in oxygen and nitrogen but lacks carbon. It suffers modulation with the solar cycle in the same sense as galactic cosmic rays, so it appears to be either galactic in origin or to be accelerated in the outer portions of the heliosphere. Its source and acceleration mechanism is a puzzle.
- Observations of discrete sources of gamma rays with energies to 10^{15} eV with ground-based detectors have identified a few cosmic-ray accelerators of great power.
- At the highest energies, above 10^{17} eV, ground-level air-shower measurements now give clear evidence of anisotropy in arrival direction; above 10^{19} eV this anisotropy suggests that these most energetic particles in nature may be of extragalactic origin.
- Large new underground detectors designed primarily to search for nucleon decay have observed and measured the flux of neutrinos from cosmic-ray interactions in the atmosphere. These detectors are also being used to study multiple muon events and their relation to the composition of primary cosmic rays around 10^{15} eV.

In addition to these discoveries, a number of other observations also raise important questions for the future. These include the following:

- Measurements of secondary products of cosmic-ray nuclear interactions in the interstellar medium indicate an energy dependence of the confinement process at energies from 1 to 100 GeV/amu (1 GeV = a billion electron volts). Unexpectedly high fluxes of antiprotons suggest that the cosmic-ray protons that produce them penetrate more matter before reaching us than do heavier cosmic rays. These data have altered our picture of the processes by which cosmic rays are confined to the galaxy and constrain models of cosmic-ray acceleration.

- Ground-level observations indicate changes in cosmic-ray composition at energies just above those reached so far by direct measurements. It appears that between 10^{14} and 10^{16} eV the cosmic rays are richer in heavy nuclei relative to protons than they are at lower energies, while at still higher energies, above 10^{17} eV, protons may again dominate.
- In 1972 measurements of attenuation in air of cosmic-ray protons up to 50 TeV (1 TeV = 1 trillion electron volts) indicated that the proton-proton cross section increases with energy. This inference was subsequently confirmed by direct accelerator measurements. More recently, results from large air showers suggest that this increase continues at least another four decades in energy.
- A series of balloon flights of emulsion chambers has observed and measured the composition and interactions of heavy nuclei of up to 10^{14} eV. In some cases the interactions produce up to 1000 secondaries.
- The flux of solar neutrinos observed appears to be significantly lower than expected from fusion processes in the Sun. This discrepancy has become one of the major unresolved issues of current astrophysics.

The following sections explore in more detail some of the topics listed above and their implications for future research.

NUCLEOSYNTHESIS

Measurements of the abundances of elements and isotopes in the solar system, as observed spectroscopically in the solar photosphere and directly in terrestrial, meteoritic, and lunar samples, have long formed the basis of our knowledge of the history of the solar system. These solar-system abundances have in turn become the benchmark for studies ranging from stellar structure and nucleosynthesis to the age and evolution of the galaxy.

Galactic cosmic rays provide a sample of material from outside the solar system, which can be used to describe the composition of the Milky Way Galaxy at a time and place far removed from solar-system formation. The cosmic-ray measurements complement spectroscopic information derived from optical and millimeter-wave astronomy on stars and the interstellar medium. Some elements and isotopes that cannot be measured well spectroscopically are relatively easy to investigate in the cosmic rays, for example, neon, iron isotopes, and many of the rare elements heavier than iron.

Abundances of radioactive nuclides and their daughters show that the solar system formed 4.6 billion years ago. Thus, the solar-system abundances have usually been taken to be representative of the interstellar medium at that time. However, recent observations of isotopic abundance anomalies in various meteoritic minerals give evidence for compositional inhomogeneity of the nebula that formed the solar system, and these observations give evidence for a significant "last minute" infusion into this nebula of products of supernova nucleosynthesis. Thus the solar-system abundances probably do not measure the present interstellar medium and may not even be completely representative of the general interstellar medium 4.6 billion years ago.

Recent cosmic-ray measurements have resolved clearly the radioactive nuclide ^{10}Be, which has a half-life of 1.6 million years. They demonstrate that the cosmic-ray nuclei that we observe today were typically accelerated about 10 million years ago, very recently when compared with the age of the solar system. Most of them reach us from distances much greater than a parsec but less than several kiloparsecs. Thus the cosmic rays sample a region that is large compared with the probable size of the protosolar nebula but probably does not extend to the center of the galaxy.

Recent models suggest that the acceleration of the bulk of cosmic rays occurs in supernova shock waves propagating through the hot interstellar gas. It thus may be that the cosmic-ray composition is more representative of the interstellar medium than is the solar-system composition.

While galactic cosmic rays provide an excellent sample of material from outside the solar system, energetic particles from the Sun, or solar cosmic rays, provide in some cases the best solar-system abundance data available. For example, the solar-system abundances of noble gas elements and their isotopic compositions, poorly determined from meteorites or from optical observations of the Sun, can best be measured in solar cosmic-ray composition studies.

The nucleosynthesis of the elements that make up the solar system has been understood as the sum of several processes. Primordial hydrogen and helium are burned in stellar interiors in a series of steps at increasing temperature and pressure, which release energy as lighter elements fuse to make heavier ones, building up eventually to elements in the iron peak. Elements heavier than nickel are principally produced by neutron capture, either slowly over periods of thousands of years in evolved stars—the (slow) s-process—or quickly in seconds during supernova explosions—the (rapid) r-process. Each nucleosynthesis process leaves a signature in relative abundances of various nuclides.

In the cosmic-ray source composition we look for signatures that reveal the conditions under which these nuclei were synthesized. We also test models of nucleosynthesis based on solar-system abundances.

Two points are clear from data already in hand: (1) The material that is accelerated to form cosmic rays has a composition that is different from that of the material that formed the solar system. This difference must reflect a difference in the conditions under which nucleosynthesis took place, or at least a different mixture of material from the various nucleosynthesis processes. (2) The composition of the cosmic-ray source material is distinguished from that of the solar system by subtle quantitative differences that require precise measurements. These points are pertinent to the plans for the next generation of experiments.

Isotope Ratios

Quantitative differences between cosmic-ray source and solar-system composition have been established by isotopic measurements with excellent mass resolution of the elements Ne, Mg, and Si. The abundance ratio ^{22}Ne/^{20}Ne is higher in the cosmic-ray source than in the solar system by a factor of about 4. The four relatively rare neutron-rich isotopes of Mg and Si are all about 60 percent more abundant in the cosmic rays (relative to the most abundant isotope of each element) than in the solar system.

Several mechanisms have been postulated to explain these cosmic-ray enrichments of the heavier isotopes. These mechanisms involve nucleosynthesis of cosmic-ray elements under different conditions from those in the solar system, owing either to spatial inhomogeneities in the galaxy or to chemical evolution of the galaxy in the time between formation of the solar system (4.6 billion years ago) and acceleration of the cosmic rays (only about 10 million years ago). These mechanisms lead to quantitative predictions for expected isotopic composition of other cosmic-ray elements so that measurements with much higher statistical accuracy than are currently available of the elements S, Ar, and Fe should be able to distinguish among various models.

Abundances of Heavy Elements

In the charge region beyond Fe and Ni, the HEAO-3 experiment has shown that the cosmic-ray source is not dominated by a single nucleosynthesis process such as the *r*- or *s*-process. However, these results do not rule out an enhancement by a factor of as much as 2 in either the *s*-process or the *r*-process contribution relative to the solar system. If the solar system were enriched in products of explosive (supernova) nucleosynthesis due to a nearby supernova shortly before condensation of the solar nebula, while the cosmic rays were a sample of "normal" interstellar material, lacking the "last minute" r-process enrichment of the solar system, then one would expect the cosmic rays to appear enriched, by perhaps a factor of 2, in s-process nuclides. Further measurements of abundance ratios of heavy elements will help to resolve such questions. Precise decomposition of cosmic rays heavier than Ni into r- and s-process components will ultimately require isotope measurements.

Both HEAO-3 and Ariel-6 data demonstrate that the abundance of actinide elements ($Z > 90$) in the cosmic-ray source is not greatly enhanced compared with that in the solar nebula, as was suggested by earlier measurements. In fact, the observed ratio of actinides to elements in the $Z = 80$ region is roughly 1 percent. This result already rules out a classical, actinide-producing r-process episode of explosive nucleosynthesis in supernovae as the source of heavy cosmic-ray nuclei. However, this actinide abundance is so low that its measurement is limited by poor statistics; only one and two actinide nuclei have been observed by HEAO-3 and Ariel-6, respectively.

Measurements of the relative abundances among individual actinide elements would show the age of these elements since nucleosynthesis. Figure 16.1 shows the expected relative abundances of actinide elements as a function of time since synthesis in an r-process event. A synthesis age of the order of 10 million years (the same as the cosmic-ray propagation time) as indicated by a U/Th ratio of about 5 would, for example, imply that cosmic-ray acceleration acts on freshly synthesized material and so would contradict the idea that the cosmic rays are a sample of today's general interstellar medium. On the other hand, if we assume that cosmic rays are a sample of today's interstellar medium and the solar system is a sample from 4.6 billion years ago, the U/Th ratio in the cosmic rays would provide a measure of the rate of r-process nucleosynthesis in the galaxy since the formation of the solar system.

Solar Neutrinos

Recently the capability of detecting neutrinos from the Sun has opened a new window on stellar nuclear processes. The nuclear fusion occurring in the Sun is calculated to produce a detectable flux of electron neutrinos, and accordingly a large-scale experiment has been operating over the past decade in a South Dakota gold mine. In this experiment the inverse beta-decay of ^{37}Cl to ^{37}A is detected as

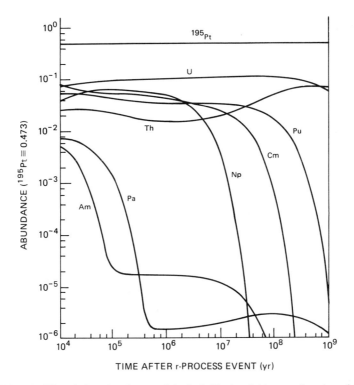

FIGURE 16.1 The relative abundances of the individual actinides as a function of time after their nucleosynthesis in an r-process event.

evidence for neutrino capture. The results of this experiment are enigmatic and important; they suggest a flux of neutrinos less than a third that calculated. As the neutrinos responsible for this reaction are of rather high energy, they come from a minor component of the solar nuclear cycle (boron beta-decay). The reason for the low flux might be due either to an error in our understanding of the solar cycle or to the loss of neutrinos through oscillations or other effects in the propagation from the Sun. In any case this experiment poses an outstanding challenge to our understanding of the astrophysics of stellar interiors, of nuclear physics, and of the elementary-particle physics of neutrinos.

ACCELERATION

Recent gamma-ray observations indicate that the bulk of the cosmic radiation of energy less than 10^{13} eV observed near Earth originates in our galaxy. Coupled with the cosmic-ray age since acceleration and an energy density outside the heliospheric cavity of 1 eV/cm^3 or greater, this suggests an average cosmic-ray luminosity close to 10^{41} ergs/s for our galaxy. This is at least 10 times greater than the x-ray luminosity of our galaxy.

Understanding galactic cosmic-ray acceleration is part of a concentrated effort to understand all classes of energetic particle acceleration in astrophysical settings. Acceleration of particles by the Sun has been directly observed. The scale of solar acceleration (energy, time, size) is much smaller than that for galactic cosmic rays. The latter can be as much as a million times more energetic than solar cosmic rays. Nevertheless some of the same theoretical approaches are used to understand both types of process. In addition, we see direct evidence via electron synchrotron emission that acceleration is also going on in such diverse objects as supernova remnants, radio galaxies, and quasars. If our experience with galactic cosmic rays is any guide, these objects may contain at least 100 times more energy in cosmic-ray nuclei. The acceleration of energetic particles is apparently a universal phenomenon and deserves a concentrated effort toward its understanding.

Shock Acceleration

Energy requirements suggest supernovae as the cosmic-ray sources, and early models of cosmic-ray origin assumed these discrete sources. The power-law spectrum led to later models, which incorporated diffuse, relatively slow acceleration by random collisions with massive moving magnetic knots in the interstellar medium. Then a trend back to discrete sources such as supernovae or pulsars took place because of the inefficiency of such second-order Fermi acceleration. This evolution of ideas has been driven by continued improvement of the observational evidence and development of the theories. The most recent acceleration models incorporate shock waves generated by supernova explosions traveling in low-density regions of hot interstellar gas, which accelerate cosmic rays trapped in the shock front.

Essentially direct observation of acceleration of particles by shock waves in the solar cavity has stimulated and guided the development of the theory of shock-wave acceleration generally. Within the solar system there is enough information to relate the shape of the spectrum of accelerated particles and its termination to the nature and size of the accelerating shock. Extending this kind of understanding to galactic scales is clearly desirable.

The most decisive observational constraints to theories of galactic acceleration will come from measurements of the energy spectra of the various cosmic-ray components; in particular, the energy dependence of the secondary/primary ratio at high energies is an important test of models of cosmic-ray acceleration and confinement. Currently available data on the composition extend only to about 10^{13} eV total energy. At still higher energies, our information at present is restricted to the study of showers of secondary particles in the atmosphere, making possible a determination of the overall energy spectrum of the parent particles but providing only an estimation of the primary composition. A better understanding of high-energy composition is essential.

Acceleration Fractionation

There is clear evidence that cosmic-ray elemental abundances after acceleration differ by factors of 2 to 10 from one element to another relative to the standard accepted solar-system abundances (derived from meteorites and the photosphere). These differences are organized, at least to first order, by atomic properties of the elements; in particular Figure 16.2 shows that there is a clear correlation between the ratio of cosmic-ray source abundance to solar-system abundance and the first ionization potential of the element. This correlation suggests that the differences are affected by fractionation in the acceleration process or in some process that injects material into the acceleration region.

A similar correlation with first ionization potential has been observed for the abundances of elements in the solar energetic particles when compared with the standard solar-system abundances, leading to the suggestion that similar fractionation effects occur in both solar and galactic acceleration or injection. An alternate viewpoint suggests that the standard solar-system abundances are in fact not correctly representative of the photosphere or of the interstellar medium. Further measurements of rare elements in the galactic cosmic rays and in the solar energetic particles may help to define the role of such fractionation in the acceleration processes.

The striking underabundance of hydrogen in cosmic rays is poorly understood and does not fit the first ionization correlation. It could reflect some property of the acceleration mechanism that depends on the charge/mass ratio (which is unity for hydrogen but less than or equal to 1/2 for other nuclei). Alternatively, it could reflect a different origin for protons (and perhaps helium).

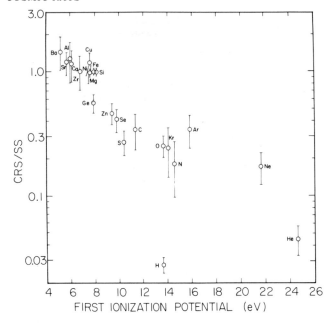

FIGURE 16.2 The elemental abundances of the cosmic-ray source relative to solar-system material are roughly ordered by the first ionization potential. However, some of the remaining differences are well beyond the indicated errors.

Termination of Acceleration Mechanism

Of particular importance in the future will be precise measurements of the proton spectrum extending to energies between 10^{13} and 10^{15} eV/nucleon. Here both the time and the size scales of the acceleration region for nuclei will eventually limit the energy attainable, leading to a break in the spectrum.

Air-shower observations (which measure the spectrum of the total energy of cosmic rays—their energy/nucleus) indicate that in the region around 10^{15}-10^{16} eV (where the spectral steepening occurs), the composition may become enriched in heavier nuclei. A rigidity-dependent termination of acceleration, as in the shock mechanism, implies a progressive enrichment in heavy nuclei with increasing energy per nucleus. It is not yet clear, however, whether this picture is correct in detail. Direct observations of the composition and spectra between 10^{13} and 10^{16} eV are required in order to understand galactic cosmic-ray acceleration and containment models.

The acceleration of solar-flare particles (solar cosmic rays) is another question. While the mean composition, averaged over many flares, is similar to that of the galactic cosmic-ray sources, including a correlation with first ionization potential, there are dramatic flare-to-flare variations that remain to be explained. The ratio of heavier elements (e.g., iron) to lighter elements (e.g., oxygen) varies by an order of magnitude from flare to flare, and the energy spectra also show wide variations. In addition some flares have anomalously high fluxes of ^{3}He, thought to be the result of a cyclotron resonance in the acceleration region. Testing models of flare acceleration require correlated observations of particle spectra and of x rays (from accelerated electrons) and gamma rays (from accelerated nuclei), as well as further measurements of the recently observed neutron flux from solar flares.

High-Energy Gamma Rays

Gamma rays of 10^{11}-10^{16} eV energy produce electromagnetic cascades in the atmosphere that can be studied from the ground using atmospheric Cerenkov emission and cosmic-ray air-shower techniques. The Cerenkov light from these air showers is almost parallel to the shower direction (to approximately 1 degree) so that a telescope image of this Cerenkov light reveals a fuzzy spot that gives the direction from which the primary cosmic ray or gamma ray arrived. Recently, experiments involving surface arrays of particle detectors have identified gamma rays of up to 10^{15} eV from Cygnus X-3 and possibly from other objects. By tracking the astronomical object of interest it is possible to separate the point-source gammas from the isotropic background of air showers produced by charged cosmic rays. The signal-to-noise ratio may be further aided through the use of accurate timing and the known timing of the source emissions.

These studies are technically only an extension of astronomy to an extreme energy of the electromagnetic spectrum. The techniques used tie this area to other cosmic-ray programs. It is noteworthy that, with these observations, our study of radiation from the universe spans over 20 orders of magnitude (10^{20}) in wavelengths of electromagnetic radiation. The results so far have given us the first direct evidence of discrete astronomical locations of acceleration processes with energies of 10^{15} eV (1000 TeV). Although this field is only a few years old, the results are already having a major impact on our understanding of the origin of cosmic rays.

Anomalous Component

An acceleration process that may have special significance, but about which very little is known, is responsible for the so-called anomalous component. Enhanced fluxes of certain nuclei such as He, N, O, and Ne are observed near the Earth at energies of 10 MeV/nucleon. Why only certain elements are enhanced, how they are accelerated, and why they appear at the Earth only at certain times are subjects of much discussion. A solar origin appears to be ruled out. We may be seeing direct selective acceleration of particles originating in the local interstellar medium, or we may be seeing particles from sources nearby in the galaxy with unusual composition. In either case, measurements of the charge and isotopic composition of these particles must be made at a new level of accuracy to understand the processes involved.

GALACTIC COSMIC-RAY TRANSPORT AND THE INTERSTELLAR MEDIUM

The cosmic-ray flux arriving near the Earth results from a convolution of source composition(s), charge-dependent selection during acceleration, fragmentation from interactions with the interstellar medium en route, and diffusion and scattering processes in the galaxy. Separating these different physical phenomena is a major task for the cosmic-ray program in the coming years.

The galactic cosmic rays constitute a highly relativistic gas held in the galaxy for a time (10^7 years) that is long compared with the traversal time for highly relativistic particles across the galaxy (10^3 years) but short compared with the age of the galaxy (10^{10} years). The physics of containment is poorly understood. We know from measurements of Faraday rotation and the polarization of starlight that the typical interstellar magnetic field is ~3 µG. Thus, galactic cosmic rays, which range in energy from 1 GeV/nucleon to greater than 10^6 GeV/nucleon, have gyroradii that range from about 0.1 AU to greater than 1 parsec (pc). However, the distribution of fluctuations in magnetic-field magnitude and direction, which are presumably responsible for scattering the cosmic rays and trapping them in the galaxy, is unknown. Current estimates suggest that the bulk of the cosmic rays diffuse to us from distances greater than a parsec but less than several kiloparsecs.

Key observational parameters for the question of cosmic-ray propagation and containment in the galaxy are the abundances of secondary cosmic rays (produced by interactions with the interstellar gas) relative to primary particles (accelerated in source regions). Particularly important are positrons and antiprotons (generated by primary protons); the light elements Li, Be, and B (fragmentation products principally of C and O), and certain heavy elements, in particular Sc and V (produced by spallation of Fe nuclei). Abundances of secondaries, together with the fragmentation cross sections, give a measure of the average path-length λ_e traversed by cosmic rays in their lifetime. If the average density of the medium is known, this can be translated into an average lifetime.

Energy Dependence of Escape from Galaxy

A fundamental result of measurement of secondary nuclei is that at higher energies the path length decreases approximately as $\lambda_e(E) \propto (E/E_0)^{-0.5}$, decreasing to $\lambda_e \sim 1$ g/cm^2 at around 100 GeV/nucleon. Only if such measurements are continued to still higher energies, i.e., well into the TeV/nucleon region, may one be able to explain the origin of this energy dependence of λ_e. For example if it is a consequence of the diffusion and convection processes by which cosmic rays are transported out of the galactic confinement volume, then λ_e is predicted to continue to decrease as energy increases at a rate that reflects the spectrum of magnetic inhomogeneities in interstellar space. If, on the other hand, the effect is due to an energy-dependent escape mechanism in regions surrounding the acceleration sites, then λ_e would become independent of energy at a value reflecting the amount of material traversed by cosmic rays after leaving the source. Several predictions are shown in Figure 16.3. It is important to emphasize that measurements above 100 GeV/nucleon will not only specify the mode of propagation of cosmic rays in the galaxy but will also enable us to deduce the energy spectra at the acceleration site.

The behavior of the escape length as a function of energy below 1 GeV/amu is a subject of considerable current interest. There is some evidence that the distribution of escape lengths is energy dependent with an energy-dependent deficiency of short path lengths. Such a path-length distribution could result from a shell of material around the source regions, in which particles are trapped in such a way that low-energy particles pass through more material before escaping than do higher-energy particles. There is also evidence that the mean escape length becomes independent of energy below about 1 GeV/amu, a

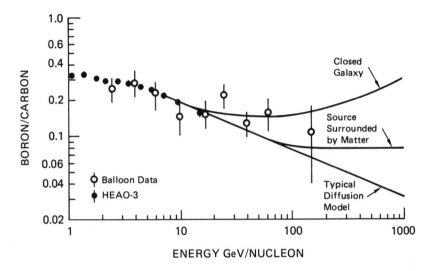

FIGURE 16.3 Various models for the containment and propagation of cosmic rays in the galactic magnetic fields will be tested by measurements in the energy range 1000 GeV/nucleon. Errors quoted for the highest-energy balloon data are much larger than those that can be obtained from satellite observations of sufficient duration.

feature that is associated in some models with a change from a high-energy diffusion-dominated transport in the galaxy to a convection-dominated regime as has been postulated in association with a galactic wind. This situation would be clarified by extending these studies to particles whose energy in the interstellar medium is below a few hundred MeV/amu, which requires direct observations of the unmodulated cosmic-ray spectra outside the solar system, or possibly over the solar poles.

The low-energy galactic cosmic rays are also of interest because they are highly ionizing and couple strongly to the ambient interstellar medium. The cosmic-ray energy density is comparable with or greater than that of the interstellar magnetic field and the turbulent motion of the gas. Cosmic-ray pressure creates bubbles in the interstellar magnetic field, puffing it out of the galactic plane, leading to the escape of cosmic rays. At the same time, gas then flows down the magnetic field, attracted by the gravitational potential of the galaxy, creating a shock wave that might trigger stellar condensation. Measurements outside the heliosphere are required to determine the contribution of these cosmic rays, most of which have energies below 100 MeV/nucleon.

Correlation Between Anisotropy and Energy

There is a striking correlation between the anisotropy and the flux of cosmic rays, as shown in Figure 16.4. If the anisotropy reflects large-scale flow patterns, a simple interpretation would suggest that there is a single underlying source spectrum of $E^{-2.47}$ all the way from 10^{12} to 10^{19} eV, with the remaining observed structure associated with failure of the containment mechanism. The anisotropy measurements are made with long-duration, ground-based experiments—observations of muons underground at the lower energies and monitoring arrival directions of extensive air showers at higher energies. Statistical uncertainties are large at the higher energies, and measurement of composition around 10^{15} eV is crucial for understanding these intriguing results.

Secondaries from Light Nuclei

A special role is played by the electron component in the high-energy cosmic rays. Cosmic-ray electrons, consisting of negatrons mostly accelerated in source regions plus positrons that are predominantly the result of interstellar p-p collisions, rapidly lose energy through radiative interactions with the interstellar magnetic and photon fields. This energy loss gives rise to much of the observed nonthermal radio and

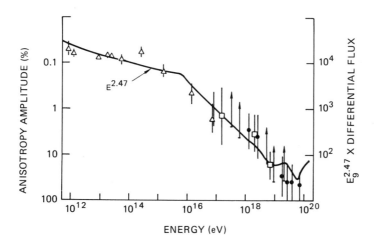

FIGURE 16.4 Amplitude of first harmonic as a measure of residence time: anisotropy (data points) compared with flux (line). Anisotropy has been corrected for solar motion below 10^{15} eV. [After A. M. Hillas, Annu. Rev. Astron. Astrophys. *22*, 425 (1984).]

x-ray background emission of the galaxy. Because of radiative losses, the lifetime of electrons, and hence the distance they can propagate in the galaxy before losing a significant fraction of their energy, decreases rapidly with increasing energy. Thus electrons, observed with an energy of a few TeV at the Earth, must have been accelerated not further than a few hundred parsecs from the solar system. Measurement of these high-energy electrons therefore provides the unique possibility of identifying the distribution of local sources of the cosmic radiation. In the past 15 years, the total electron flux has been measured to about 1 TeV.

The observation of the energy spectrum of positrons has a special importance. It makes possible a direct comparison to the source spectrum of positrons, which is known through calculations of the p-p production process measured at accelerators. At present the positron spectrum is known separately only to around 10 GeV. If this measurement could be continued up to a few hundred GeV, it would give direct information on the deformation of the spectrum due to propagation effects and radiative energy losses. Such information cannot be unambiguously obtained just from observations of electrons since their energy spectrum at the source is not known *a priori*. Thus, positron observations would lead to independent determinations of the confinement time of the electron component in the galaxy together with an estimate of the magnitude of the magnetic field traversed.

Observations of other kinds of secondaries such as antiprotons, ^{2}H and ^{3}He from interactions of protons, and helium nuclei provide information on the amount of matter traversed by the most abundant cosmic-ray constituents. Recent measurements of relatively high antiproton intensities at around 10 GeV suggest that protons may traverse 3 to 5 times as much matter as heavier nuclei. A similar situation seems to exist for helium based on recent observations of a high ^{3}He/^{4}He ratio. Very-low-energy antiproton measurements are even more difficult to interpret. More accurate observations of positrons and antiprotons at different energies and of deuterium and ^{3}He should be able to decide the question of whether protons and helium nuclei have different propagation histories from those of heavier nuclei.

Propagation in Galactic Halo

Observations of the radioactive secondary nucleus ^{10}Be, interpreted within a simple (leaky-box) propagation model, indicate a cosmic-ray lifetime of about 10 million to 20 million years. Comparison with the average path length deduced from the secondary/primary ratio mentioned above implies that the cosmic rays observed at the Earth propagate in a region with an average density less than that of the average interstellar medium in the disk. This in turn suggests a containment volume that includes a galactic halo region as well as the disk. The interrelationship between the matter traversed by the particles and their age is dependent on the size of the storage volume for cosmic rays. What is actually measured is the fraction of ^{10}Be that survives radioactive decay. This depends not only on the mean cosmic-ray age but also on the distribution of ages, which is exponential in the leaky-box model but is more complicated in models in which cosmic rays are stored in a large halo surrounding the galaxy. Further information about cosmic-ray time scales and hence about the storage volume will come from measurements of ^{10}Be abundances at higher energies and of other clock isotopes. These data, in conjunction with electron and positron measurements, would be able to differentiate between halo and local storage models and to place constraints on the distribution of cosmic-ray sources in the galaxy.

Connection with Gamma and Radio Astronomy

The cosmic-ray composition studies discussed above give information on the distribution of cosmic rays and matter in the galaxy that is complementary to that obtained with gamma-ray and radio-astronomy surveys. Diffuse gamma rays are generated by interactions of cosmic rays with the interstellar gas; the nonthermal radio emission comes from cosmic-ray electron synchrotron emission in the galactic magnetic fields. By studying this radiation we can also observe the cosmic rays in localized galactic objects (supernova remnants) and in external galaxies. These two different perspectives will be helpful in understanding the role that cosmic rays and the magnetic fields play in the evolution and dynamics of astrophysical objects, from supernova remnants to giant radio galaxies.

HIGH-ENERGY NUCLEAR AND PARTICLE PHYSICS

From the point of view of high-energy physics there are several reasons to study cosmic rays: (1) to explore particle interactions at energies much higher than those accessible at accelerators; (2) to study processes involving neutrinos and high-energy nuclei that are also inaccessible to present machines; and (3) to look for signals from the early universe, such as a cutoff of cosmic rays above 10^{20} eV due to the 3-K blackbody radiation, or the presence of antinuclei, which bears on the question of whether the universe is baryon symmetric on the largest scales. In addition there is considerable scope for applying particle physics to the study of cosmic-ray astrophysics, i.e., to determine the chemical composition and energy spectra of the primary cosmic rays in the high-energy region where the flux is too low for direct observation of the primaries.

Different types of experiments are suited to the different regions of the primary energy spectrum as determined by the flux. This is indicated in Figure 15.1(B) in Chapter 15, which shows the integral flux as a function of primary energy. A scale showing equivalent nucleon-nucleon center-of-mass energies is superimposed. Note that the region of the second-generation hadron colliders (one of which is already in operation) to a large extent overlaps the 10^{14}-10^{16} eV region, which includes the astrophysically interesting region of the energy spectrum referred to earlier.

Because of the steeply falling primary spectrum there is a natural dividing line around 10^{15} eV (or somewhat lower) between direct and indirect experiments. The total flux above this energy is only about 2 particles per (m^2 sr week) at the top of the atmosphere. Since the flux decreases by about 2 orders of magnitude per decade increase in energy, it will continue to be necessary to explore higher energies with indirect, ground-based cascade experiments. Because of the anticipated direct measurements of primary composition to 10^{14}-10^{15} eV, coupled with current studies of hadron collisions in the same energy region, there is now a good prospect for improving significantly our ability to interpret the cascade measurements at the higher energies.

Nucleon Decay Experiments as Cosmic-Ray Detectors

Motivated by the particle-physics prediction of spontaneous decay of the free (or bound) proton, large detectors have been designed and built in this country and abroad that are sensitive to nucleon decay lifetimes of as great as 10^{33} years. These large detectors represent a unique opportunity to collect data on energetic muons and neutrinos from cosmic rays. The characteristics of the U.S. detectors are noted here together with specific comments on appropriate cosmic-ray observations and opportunities.

The largest operating proton-decay experiment employs an 8000-m^3 volume of water located at a depth of 600 m, or 1570 m.w.e. (meters water equivalent), in a salt mine near Cleveland, Ohio. Signals from Cerenkov light produced by relativistic charged particles are detected by photomultipliers that line the six surfaces of the tank on a 1-m grid.

Cosmic-ray neutrino interactions depositing energies of over 200 MeV are detected at a rate of about one per day. The detector has been in operation since August 1982.

Two other smaller proton-decay experimental programs have also been carried out in the United States. At Park City, Utah, a 780-m^3 water Cerenkov detector was operated at a depth of about 1700 m.w.e. A 30-ton detector at the Soudan mine in northern Minnesota is at a depth of 1800 m.w.e. and consists of a taconite-loaded cement with proportional chambers as the sensitive elements. Although much smaller than the other two detectors, its fine-grained tracking capability has enabled the detector to search for possible sidereal anisotropies of cosmic-ray multiple-muon events. A larger detector, Soudan II, is scheduled to be constructed in the same mine employing the same general design philosophy.

An unusual experiment has been developed in the Homestake gold mine in South Dakota. This detector consists of an array of plastic tanks filled with liquid scintillator, which, when brought into full operation, will have a sensitive mass of about 300 tons. It is located in the deep underground cavern occupied by the solar neutrino experiments. The primary objective of the experiment is to search for neutrino bursts that could be signatures of supernova explosions. This counter array is, of course, also sensitive to cosmic-ray muons. A surface array is being added to study the air showers produced by the same primary events that give rise to the detected muons. Although the expected number of energetic muons increases with primary energy, at fixed energy the muon multiplicity is correlated with the atomic weight of the primary cosmic ray. Consequently, the surface shower data and underground muon data together provide information concerning the atomic weight of the primary cosmic-ray nucleus. The Homestake data will be useful in studies of the mass spectrum of primary cosmic rays in the energy range 10^{14}-10^{16} eV; these energies are about an order of magnitude greater than those accessible with the Cleveland proton-decay detector owing to the greater depth of 1480 m (4200 m.w.e.) of the Homestake mine.

Nucleus-Nucleus Collisions

The classic cosmic-ray emulsion technique has been modified into a hybrid emulsion chamber with target material and electromagnetic calorimeter sections (layers of plastic and lead, respectively, between photosensitive layers). The first observation of charmed particles was made over 10 years ago with such detectors, and they have been adapted for use at accelerators to study charmed-particle spectroscopy and lifetimes. Scientists are currently collaborating internationally on the use of such emulsion chambers supplemented by electronic detectors to study primary nuclear composition and properties of nucleus-nucleus collisions. Several balloon flights have been carried out with emulsion chamber payloads to explore primary cosmic rays in the 10^{12}-10^{15} eV energy range. This energy range is well beyond that accessible to current heavy-ion accelerators, and there are fundamental and novel questions accessible to this kind of cosmic-ray experiment, in particular, the question of whether a new phase of quark-gluon matter can be achieved in collisions between heavy nuclei at high energy. Events in which heavy cosmic rays interact to produce nearly 1000 secondary particles have been observed. The energy-density implied by such multiplicities has been calculated to be above the threshold for production of a quark-gluon phase. Over 200 interactions have been analyzed wherein the primary energy exceeds 10^{12} eV.

Cross Sections, Spectra, Anisotropies, and Composition of Primary Cosmic Rays Above 10^{17} Electron Volts

Above 10^{16} eV cosmic rays remain of interest for high-energy physicists as well as for astrophysicists, at least until the operation of a supercollider, which may be completed in the 1990s. The goal here is to determine both cross sections for hadron interactions and the composition of the primaries. Recent measurements at the CERN pp collider have confirmed earlier cosmic-ray estimates of the proton cross section up to 10^{14} eV (equivalent to center-of-mass energy of 500 GeV). New air-shower experiments have the potential to measure the proton cross section and to determine the gross features of the primary composition as well as in the 10^{17}-eV to 10^{19}-eV (center of mass about 100,000 GeV) range, where there may be a transition to extragalactic cosmic rays.

The most ambitious cosmic-ray air-shower experiment in the United States is the Fly's Eye experiment being carried out in Utah. This detector consists of two arrays of photomultipliers deployed 3 km apart to observe the air scintillation light produced by extensive air showers. The phototubes are grouped in the focal plane of spherical mirrors, so that the arrays provide a mosaic image of the sky, with each phototube sensitive to a hexagonal cone of 5° of the celestial sphere. Timing information is also available, so that an air shower is recorded as a series of phototube "hits," with a pulse amplitude and relative time recorded for each. The data are sufficient to reconstruct completely the air shower in space and absolute magnitude. The Fly's Eye data have two major strengths. First, the Fly's Eye covers or "sees" an effective area comparable with the largest surface air-shower array; the current detector is sensitive over an area of almost 100 km^2, although data can only be collected on clear, dark nights. Second, this detector permits the observation of the longitudinal profile of the shower, hence providing information on the height of the primary interaction and on the rate of development of the shower. These data in turn may be interpreted in terms of the inelastic cross section of protons at very high energies and in terms of the primary nuclear-mass composition. It may also be possible to relate the rate of development and shape of the shower with the secondary-particle multiplicity and other inclusive parameters of proton interactions.

The Fly's Eye experiment has achieved a major milestone by directly observing the longitudinal development of individual cascades. Present results from this experiment and other air-shower experiments already suggest that the proton-air cross section is larger than 500 mb at 10^{18} eV, as compared with its low-energy value of 280 mb.

Magnetic Monopoles

Most Grand Unified Theories (GUTs) predict the existence of massive magnetic monopoles, quanta of isolated north or south magnetic poles with discrete magnetic-pole strength. Their masses are predicted to be of the order of 10^{16} GeV (or about 0.01 µg), although some models yield significantly lighter or heavier masses. In the standard big-bang cosmology, GUT monopoles are produced at an early stage of the universe. By contrast, in the inflationary-universe scenario there would be no significant monopole production.

The density of monopoles in the universe today can be bounded by arguments based on the openness of the universe and the mass of missing, or dark, matter. Another astrophysical upper limit on the monopole flux is based on the long-term stability of galactic magnetic fields. Within these limits, monopoles may exist in the universe with velocities in the range of 10^{-4} to 10^{-3} the velocity of light. At the lower end of this velocity range, some theorists suggest that they could be gravitationally bound to the solar system, which might enhance their local abundances.

If GUT monopoles are able to catalyze proton decay, as suggested by some current theories, they would produce copious x rays from neutron stars. Our present failure to observe these x rays can be used to set more stringent limits to the monopole flux for this specific monopole type. The proton-decay catalysis would also be detectable in proton-decay experiments; thus far this process is not observed.

Searches for monopoles have been conducted with superconducting coils and with ionization and scintillation detectors. In the former, a monopole passing through a coil would induce a current step that is readily detectable with sophisticated instrumentation. This technique has the advantage that the monopole signal would be almost totally independent of the monopole velocity. Such coils are limited, however, in their size. Ionization and scintillation detectors can be made with larger areas but are calculated to be insensitive to monopoles moving slower than about 5×10^{-4} the velocity of light.

A signal consistent with a monopole interpretation was reported in early 1982, using a superconducting coil. However, subsequent searches by three groups (including the original 1982 author) have failed to find further evidence for a monopole using the same technique. These searches have extended the sensitivity by almost a factor of 100. In addition, data using scintillators have set still more stringent limits on the flux over the velocity range accessible to them. Although the 1982 event remains unexplained, the monopole hypothesis for that event now seems unlikely.

17

Opportunities

In describing the opportunities for progress in cosmic-ray physics it has been convenient to consider separately those areas requiring measurements above the atmosphere, either on satellites or on stratospheric balloons, and those areas using earthbound instruments, either on or under the surface. Spaceborne instruments measure directly the charge, energy, and in some cases the mass of individual cosmic-ray particles with energies up to about 10^{14} eV. Ground-based instruments infer energy spectra and composition of cosmic rays above about 10^{14} eV by measurements of the showers of secondary particles produced by interactions of the primary cosmic rays in the atmosphere. In addition, for particles that penetrate the atmosphere, such as neutrinos and perhaps magnetic monopoles, certain ground-based instruments may detect the primary particle. There also has been a practical, organizational difference between ground-based and spaceborne measurements. The spaceborne measurements have been funded principally by the National Aeronautics and Space Administration (NASA), while the ground-based experiments have been funded primarily by the National Science Foundation and the Department of Energy.

SPACEBORNE EXPERIMENTS

Developments of both spacecraft and instrumentation over the past decade, combined with NASA's plans for a Space Station in the early

1990s, provide us with opportunities for definitive cosmic-ray experiments in space. There are a number of important measurements that can be made with a superconducting magnetic spectrometer facility on the Space Station. There are also important observations that can be made with instruments already built or under construction, attached to the Space Shuttle, the Space Station, or the Long Duration Exposure Facility. In addition there are also a few key experiments, using space-proven solid-state detector technology, that require exposure outside the magnetosphere and can be placed there with Shuttle launch and subsequent upper-stage boost.

In this section we describe scientific questions that can be answered in the next decade with these existing technologies, and in the recommendations that follow we again emphasize the next decade and experiments that we now know to be feasible. In a longer-term view there have been suggestions for assembling in space much larger cosmic-ray experiments capable of extending our knowledge even further. Undoubtably these further developments should be studied in the next several years.

Isotopes

GALACTIC COSMIC-RAY ISOTOPES

We now have in hand techniques to measure the mass of individual nuclei and thus determine the isotopic composition of cosmic-ray elements. With the discovery that the heavy stable isotopes of Ne, Mg, and Si are enhanced in galactic cosmic rays, it appears highly likely that the isotopic composition of other less-abundant elements will also differ from that of solar-system material because of different nucleosynthetic history. The measurements of the neutron-rich isotopes of S, Ar, and Ca, for example, are required to distinguish among models that have been proposed to explain the Ne, Mg, and Si abundance anomalies. But lower fluxes and larger contributions from interstellar fragmentation of heavier elements require significantly larger instruments to be able to gather adequate numbers of nuclei to make definitive conclusions.

Using well-established techniques of solid-state detectors, it is now possible to construct an instrument with sufficient mass resolution and size that with a few-year exposure outside the magnetosphere the detailed isotopic composition would be determined at energies of a few hundred MeV/amu for all elements in the galactic cosmic rays up to atomic number 30. Such an instrument could be flown on an Explorer-class mission.

With a superconducting magnetic spectrometer on the Space Station, isotope measurements of similar precision and sensitivity would be possible at much higher energies—several GeV/amu. This extension of isotope measurements to energies where such measurements were previously impossible will permit probing of a variety of cosmic-ray time scales using radioactive isotopes at large Lorentz factors and will probe for energy dependence of sites of cosmic-ray acceleration by comparison of the isotope compositions at various energies. With developments over the past decade in superconducting magnet technology in a wide variety of ground-based applications, and the developments of cryogenic applications in space, such a device appears to be quite feasible for installation on the Space Station in the early 1990s.

SOLAR-FLARE ISOTOPES

We have only limited direct knowledge of the Sun's elemental composition and almost no direct knowledge of its isotopic composition. Spectroscopic measurements of solar isotopes are difficult to perform; there are observations for only a few of the first 30 elements, and the uncertainties are large. Recently, unexpected isotopic anomalies in a number of elements have been discovered in meteorites, giving evidence for the inhomogeneity of the solar system at the time of its formation, and perhaps for nucleosynthesis activity in the solar neighborhood immediately before the formation of the solar system.

Recent measurements of solar-flare particles with cosmic-ray instruments have shown that neon in solar flares has a different isotopic composition than neon in the solar wind but the same isotopic composition found in some meteoritic components. Other heavy elements for which solar-flare isotope observations have been made show no anomalies at the 30 percent level, but observations with much better statistics are needed to determine composition at the level at which meteorite anomalies are observed—a few percent or less.

The same spacecraft outside the magnetosphere described above for galactic cosmic-ray measurements at a few hundred MeV/amu can also carry a similar instrument to measure the isotopic composition of solar-flare particles above about 5 MeV/amu.

Ultraheavy Elements

The quantitative study of ultraheavy (atomic number greater than 30) nuclei has begun with the HEAO-3 satellite. Individual element abundances have been measured for elements of even atomic number up to about 60. At higher atomic numbers, resolution and statistics limited the quality of the results to relative abundances of charge groups. For the actinide elements, around atomic number 90, the quality of the results is limited by extremely low statistics; a total of only three actinide nuclei were identified in the two experiments.

With sufficient improvements in both statistics and charge resolution, significant new results can be expected. For example, with a 2-year exposure of a 100-m^2 sr detector one could look for specific elemental tracers of recent r-process nucleosynthesis such as ^{93}Np, ^{94}Pu, and ^{96}Cm. If the fraction of r-process material were appreciably greater than 10 percent, one could even estimate the time of the r-process addition from the relative abundances of these elements.

It appears that this next major step in the study of ultraheavy nuclei can be achieved relatively inexpensively using newly developed plastic track detectors on a flight of the Long Duration Exposure Facility (LDEF), a large nearly completely passive spacecraft. The requisite number of nuclei can be detected with this large system, and it appears that sufficient charge resolution can be achieved with proper attention to temperature control and monitoring. Construction of this instrument has begun, in preparation for a launch in 1987.

High-Energy Composition and Spectra

The energy spectrum of protons, the most abundant cosmic-ray species, has been reasonably well measured on balloons up to energies around 1000 GeV, and measurements of helium nuclei exist up to more than a few hundred GeV/amu. Information on the more abundant of the heavier nuclei (carbon, oxygen, and iron) exists up to about 100 GeV/amu, although the statistical accuracy of the data is still limited. Relative abundances of the secondary nuclei that result from interstellar spallation are quite well measured at energies up to about 20 GeV/amu; spectra of these elements define the galactic confinement and propagation of cosmic rays.

Various models have been developed for the acceleration of primary cosmic rays and the production of secondaries during propagation. These models have been constructed to agree with the observed data up to about 100 GeV/amu but make different predictions about the spectra at higher energies. Precise observations at very high energies are, therefore, crucial to distinguish among these models. Ground-based measurements have great difficulty in distinguishing individual cosmic-ray elements, so it is necessary to make direct measurements in space, but the low fluxes of these higher-energy cosmic rays require large instruments and long exposures.

A large-area instrument designed to measure the energy spectra of cosmic rays with atomic number 3 through 28 at energies up to a few TeV/amu was successfully flown on the Spacelab for a week in August 1985. The fluxes of very energetic cosmic-ray nuclei are extremely low, so a 1-week exposure gives results that are limited by statistics. Reflight of this instrument on a later Spacelab mission would thus be valuable. Furthermore, attaching this instrument to the Space Station for a year would permit it to extend measurement another decade in energy, approaching the region where inferences from ground-based air-shower detectors suggest a change in the cosmic-ray composition.

Complementary observations, with much better energy resolution but at not quite so high energies, up to several hundred GeV/amu, would be possible with a superconducting magnetic spectrometer facility on the Space Station. These observations would permit, for the first time, measurements of fine structure in the cosmic-ray energy spectrum, which might be expected from a superposition of sources with different energy spectra.

Positrons, Antiprotons, Deuterium, and ^{3}He

Several significant questions about the galactic containment of cosmic rays require the observation of the secondary cosmic rays generated by interstellar collisions of the most abundant cosmic-ray species—protons and alpha particles. These observations are best performed with counter telescopes featuring magnetic spectrometers with superconducting magnets flown on the Space Station.

Measurements of the positron-to-antiproton ratio can be used to determine the critical energy at which radiative losses dominate escape losses (because both positrons and antiprotons are produced in the same collisions, but the positrons have significant radiative energy losses). This critical energy is related to the root-mean-square transverse magnetic field and to the containment time in the storage region. In order to determine this critical energy the positron-to-antiproton ratios must be measured at least up to 100 GeV. Such an exposure is possible on a 5- to 10-day Spacelab mission.

Observations of deuterium and ^{3}He are aimed at determining if helium has the same acceleration and propagation history as heavier nuclei and at making detailed measurements of the energy dependence of the confinement time in the galaxy. The first of these objectives can be met with measurements in the 1-10 GeV/amu range on 1- or 2-day flights of high-altitude balloons or in 1-week Spacelab flights. The second objective requires measurements up to about 150 GeV/amu and requires longer spaceflight exposures as would be afforded by a superconducting magnetic spectrometer on the Space Station.

Antimatter

One of the most fundamental questions in cosmology is the symmetry or asymmetry between matter and antimatter in the universe. While current cosmological models favor an asymmetry, experimental limits on extragalactic antimatter are inconclusive. Current limits on the presence of antimatter of heavy nuclei in the cosmic rays are at the level of parts in 10^4. If distant clusters of galaxies composed entirely of antimatter exist, they may contribute to the cosmic-ray flux in our galaxy at a level of at most 10^{-7} or 10^{-6}. A search at this improved sensitivity level is therefore meaningful. For an antihelium search, the required number of events could be achieved with an exposure of 0.2 m^2 sr day, attainable with a Shuttleborne superconducting spectrometer. For anti-iron nuclei, plastic track detectors in combination with plastic scintillators have been proposed, making use of the differences in energy-loss mechanisms in the two kinds of detector, with the differences depending on charge-cubed terms in the collision cross section of nuclei with atomic electrons; here too the necessary exposure could be attained with several balloon flights or a 1- to 2-week Shuttle flight. A much more sensitive search, at the level of 10^{-8}, over the full range of abundant elements from helium through iron, would be possible with a superconducting magnetic spectrometer on the Space Station.

Nucleus-Nucleus Interactions

The study of nucleus-nucleus interactions at high energies has become of great interest in the past few years because of elementary-particle theories that predict new states of matter that can be created only in such collisions, in particular the quark-gluon plasma. In addition information about nucleus-nucleus as well as proton-nucleus collisions is required for interpretations of air-shower data. At present nucleus-nucleus interactions at energies above 4 GeV/amu can be studied only in the cosmic rays. Such studies using emulsion-chamber techniques can also give the composition of the cosmic rays causing these interactions. Balloonborne exposures of such detectors have begun to make significant contributions in this field, and with extended exposures on balloons and on the Space Shuttle we can expect a significant increase in both the number of interactions studied and the highest energies observed.

Solar Modulation of Cosmic Rays

The continuous radial flow of coronal plasma and magnetic field outward from the Sun results in a cosmic-ray flux in the inner solar system that is significantly lower than in interstellar space. This effect is significant at energies below several GeV/amu, and the effect increases at lower energies. Indeed interstellar cosmic rays below a few hundred MeV/amu cannot reach the inner solar system at all, at least not near the ecliptic plane. The particles observed near the Earth below this energy had higher energies when they were outside the solar system. The magnitude of this solar modulation varies substantially and rather irregularly during the 22-year solar cycle.

The deep-space probes Pioneer 10 and 11 and Voyager 1 and 2, which are leaving the solar system, are providing important data on the extent of the modulating region. The Ulysses spacecraft, which will be launched in 1986 and will fly over the pole of the Sun at a distance of about 1 AU, will provide a direct measurement of modulation effects in a region of the solar system where the interplanetary magnetic field has a configuration different from that near the ecliptic plane. As these probes penetrate uncharted regions of the solar system, it is important to preserve monitors of the magnitude of the solar modulation including near-Earth spacecraft and ground-level neutron monitors. Neutron monitors provide a precise continuous monitor of the cosmic-ray flux at the Earth by measuring secondary nucleons produced in the atmosphere by nuclear interactions of primary cosmic-ray nuclei. A base of nearly 40 years of continuous observations is available for intercomparison of observations made at different times in the solar cycle.

GROUND-BASED EXPERIMENTS

Gamma-Ray Astronomy

The opportunity to observe directly sources of very energetic particles opens a new frontier in astronomy and astrophysics. The detection of gamma rays of over 10^{12} eV from the ground using optical Cerenkov light (10^{11}-10^{13} eV) or using extensive air-shower counter arrays (10^{13}-10^{16} eV) is at present one of the most exciting and rapidly developing fields in cosmic rays because of the potential for observing high-energy natural accelerators at work. Recent discoveries indicate that a number of binary x-ray sources, such as Cygnus X-3, Vela X-1, and LMC X-4, are sources of very energetic cosmic rays. Indeed, Cygnus X-3 alone may be sufficient to supply all the galactic cosmic rays with energies of 10^{16}-10^{17} eV. New and better measurements are urgently needed to clarify the nature of the signals above 1 TeV from point sources and to understand their implications. Several experiments are currently being developed with this aim, and this effort deserves the strongest possible support.

Air-Shower Detectors

The only major U.S. program directed toward the study of extensive air showers produced by primary cosmic rays of over 10^{17} eV is the Fly's Eye installation in Utah, described earlier.

This detector has been expanded by increasing the number of mirrors and phototubes at the second, newer site by a factor of 3. In the future the group has plans to improve resolution and sensitivity by constructing a second-generation system using a larger number of smaller phototubes and to include optical filters to reduce the background from Cerenkov light.

There is serious discussion on the development of muon detectors and/or a surface air-shower counter array in conjunction with the Fly's Eye. Detecting the same event with both techniques would provide critical intercalibration of the Fly's Eye data with other surface-array experiments. In addition, data on the lateral spread of cascades determined from the surface array could be correlated with the longitudinal development as seen with the Fly's Eye. The surface array would also collect data during the day. This would add to the global data set on the highest-energy cosmic rays with more conventional surface-array data. As noted earlier, the spectrum, anisotropy, and composition of primary cosmic rays above 10^{18} eV are all of significant interest. Although this information is indirect and interpretation of particle physics parameters is complicated by the mixed primary composition, *there can be no other access* to this extreme energy domain above 10^{18} eV through the end of this century.

There is an inevitable quest for data beyond our present horizon of about 10^{20} eV. The rates above that energy are so low—less than one/100 km^2 year) that it is not known at this time whether the spectrum truncates or flattens out above 10^{20} eV.

Neutrino Astronomy

The proton decay detectors are able to study neutrinos as a consequence of their large detector volumes. Thus far the observed neutrino interactions are from muon- and pion-decay neutrinos, which come in turn from cosmic-ray interactions in the Earth's atmosphere. Neutrinos, like gamma rays, are unaffected by galactic magnetic fields. Further, they uniquely can penetrate all interstellar environments. Thus, it has been tempting to consider developing a neutrino astronomy to seek signals from a variety of astronomical sources. It has been proposed to instrument a large volume of seawater with photomultipliers to seek Cerenkov signals from such neutrino interactions. Such a system has been christened DUMAND for Deep Underwater Muon and Neutrino Detector. A complete DUMAND installation would contain a three-dimensional matrix of phototubes deployed to observe the Cerenkov light from energetic particles and interactions in 30 million tons of seawater under a shield of 3 to 5 km of ocean. Besides neutrino interactions, cosmic-ray muons would also be observed, and their interactions at energies in excess of those available at current accelerators could be accessible to study. Multiple muon studies, as they bear on primary composition, may also merit attention.

The Homestake detector is able to study low-energy neutrinos (a few MeV) and is sensitive to supernova processes that are predicted to produce neutrinos as a consequence of gravitational collapse. The current U.S. proton-decay detectors are primarily concerned with neutrinos expected to result from cosmic-ray interactions in the atmosphere where the observed neutrino energies range from 200 MeV to several GeV. DUMAND would focus on neutrinos of above a few hundred GeV. In its most ambitious manifestation it could have a sensitivity in principle comparable with the sensitivity of the air-shower gamma detectors discussed above. If the observed sources of gammas also produce neutrinos, one will conclude that both come from interactions of very-high-energy protons—the gammas from π^0 decay and the neutrinos from $\pi^\pm$ decay. If neutrinos are not seen, it would suggest that the gammas arise from electromagnetic processes such as bremsstrahlung and synchrotron radiation. There may also be situations where gammas are absorbed or attenuated near a source while the neutrinos are not.

The many technical problems in transforming a large volume of the ocean into a particle detector have been studied for some time. There are currently plans to proceed with a one-dimension test of a design concept. This will be a single cable containing several phototube modules along its length. It will be lowered into the ocean and operated to detect cosmic-ray muons. The results of this test will significantly influence future planning in this area.

A new proposal for a joint U.S.-Italian experiment in the Gran Sasso Tunnel in Italy has been developed. Dubbed MACRO (Monopole and Cosmic Ray Observatory), its dual objectives are monopole detection beyond the Parker limit and high-energy neutrino astronomy.

The Fly's Eye may also serve as a neutrino detector, and the group working with the detector has searched their data for upward-going air showers that would be evidence for energetic neutrino interactions. As running time accumulates, this aspect of the Fly's Eye data could take on astronomical importance. By observing very energetic upward-going showers produced by neutrino interactions in the crust of the Earth, the Fly's Eye may be able to detect neutrinos from primary protons of over 10^{20} eV interacting with the 3-K blackbody radiation.

Some perspective on the energy ranges and particle types studied with ground-based cosmic-ray experiments are summarized in the bar chart of Figure 17.1. In general, the lower limits are set by experimental techniques and the upper limits by falling fluxes.

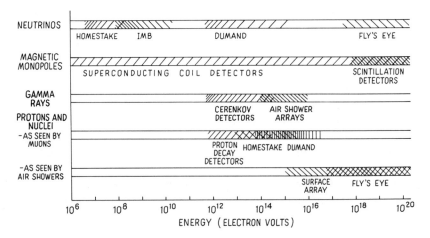

FIGURE 17.1 The range of energy sensitivity for different particle types of the present and proposed ground-based detectors discussed in this section. The lower limit is generally set by the characteristics of the detector and the upper limit by the falling spectrum of the cosmic-ray flux. The magnetic monopole sensitivities indicated are not related to flux estimates.

Magnetic Monopoles

The definitive observation of magnetic monopoles would have a major impact on our understanding of particle physics and astrophysics. This particle is so exotic and its discovery would be so important that a significant search effort is warranted.

It is possible to design much larger magnetic flux detectors—superconducting coils—than have been used up to the present. To date coil areas are 100-2000 cm^2. Groups that have operated these detectors have developed concepts for coils with a sensitive area of the order of 100 m^2, which would permit much more sensitive searches. Scintillation-counter groups have also designed large-area experiments, and at least one is in an advanced stage of construction. In addition, the Homestake detector would be sensitive to monopoles. From Figure 17.1, the interesting astrophysical upper limit to monopole fluxes is 5×10^{-16} $(cm^2 \ sr \ s)^{-1}$. To detect one event at this limit requires a detector of 1000 m^2 operating for a year. Experiments of 1000-10,000 m^2 area are possible, and detectors such as MACRO are currently being proposed that are capable of reaching this limit.

It appears that physicists will press to extend the search for monopoles until their existence is definitely confirmed or until they are not found in more than one detector of at least 1000 m^2 operating for over a year.

Nucleon Decay Detectors

The nucleon decay detectors present an unusual opportunity for cosmic-ray research. They are large underground detectors sensitive to energetic cosmic-ray muons and to neutrino interactions. The largest nucleon decay detectors have measured for the first time the flux of cosmic-ray neutrinos by directly observing their interactions inside the detector volume in significant numbers. Rates are consistent with expectation, and they offer the possibility of extending the search for neutrino oscillation by comparing fluxes of upward and downward neutrinos. If neutrinos have masses, then electron and muon neutrinos may oscillate into each other (or into other types of neutrinos) over large distances. The diameter of the Earth is so much larger than laboratory scales that this geophysical type of experiment could possibly see effects not accessible in the laboratory; mass difference down to 10^{-2} eV can be studied, although the limits on the relevant parameter, $\sin^2 \theta$, will be weaker than for laboratory experiments. At the same time, measurement of the neutrino flux and comparison with conventional expectations are being used to help calibrate the detectors for their primary mission of searching for nucleon decay.

All nucleon decay detectors observe multiple muons with varying degrees of spatial and angular resolution. If the energy per nucleon is sufficiently above threshold for production of muons in the atmosphere that can survive to the depth of the detector, then multiple-muon detection rates are in principle sensitive to primary composition because a heavy nucleus is more likely to produce a multiple-muon event than is a proton primary. Detectors now operating have already begun to collect multiple muon events with larger collection areas and a larger range of depths than has been possible previously. Sensitivity to primary cosmic-ray composition may be enhanced significantly by a surface-detector air-shower array in coincidence to estimate the energy of the primary by its accompanying shower.

Additionally, data on the lateral spacing of muons—the decoherence curve—is relevant to the transverse momentum distributions of muons and their parent pions from the primary cosmic-ray interaction in the atmosphere. As the primary cosmic-ray energies explored are within the range of the current generation of $\bar{p}p$ colliders, these data are of interest principally in the context of the properties of nucleus-nucleus collisions.

Solar Neutrinos

The importance of the solar neutrino experiment has been correctly emphasized in the report of the Astronomy Survey Committee (*Astronomy and Astrophysics for the 1980's*, Volume 1, National Academy Press, Washington, D.C., 1982). That report (page 114) ". . . recommends continued, vigorous support for programs to detect and measure the flux of neutrinos from the Sun. Additional facilities are needed to supplement the data currently being obtained by ^{37}Cl detectors. . . ." As emphasized in that report, it is feasible to use the inverse beta decay of gallium as a detector of low-energy neutrinos, i.e., the neutrinos from the proton-capture processes in the Sun. Such an experiment, although expensive, would be an independent and more definitive probe of solar nucleosynthesis.

Another interesting possibility for solar neutrino study has been discussed. It appears feasible to increase the sensitivity of the water proton-decay detectors through the addition of more and/or larger phototubes to observe solar neutrinos in real time, possibly including directional information. There could be serious backgrounds; nevertheless, if this goal could be realized, it would be one of the important opportunities of this decade.

Future Opportunities

This field of science has provided opportunities for bold, creative ideas in the past, and we should be alert to new opportunities presented by new ideas, unexpected results, or developments in related areas of physics. A second-generation Fly's Eye, a surface array at Fly's Eye, developments for neutrino astronomy including solar neutrinos, large-scale monopole detectors, MACRO, new gamma-ray detector systems, and DUMAND are potential candidate programs. Other programs involving international collaborations may also develop in the air-shower field. For example, accessible mountain-top observatories in the Andes and the Himalayas exceed by thousands of feet in elevation (therefore by one or two nuclear interaction mean free paths) any potential U.S. sites. International collaborations at these unique sites involving U.S. participation may evolve in the future. Any of these future possibilities should be regarded as serious candidates for an incremental increase in the support level of ground-based cosmic-ray experiments.

THEORY

Theoretical calculations are a vital component of cosmic-ray physics. Calculations of stellar and explosive nucleosynthesis form the basis for drawing implications about the relative importance of various astrophysical processes from measurements of cosmic-ray composition; calculation of the neutrino spectrum from gravitational collapse is closely related. Understanding the nearest star depends on modeling of the nuclear reaction cycle in the Sun and other solar calculations, which underlie, for example, interpretations of solar neutrino experiments. The processes by which cosmic rays are accelerated to exceedingly high, suprathermal energies are intrinsically interesting, and significant theoretical progress is being made in understanding them. Calculations of cosmic-ray propagation lead to understanding of the interstellar environment as well as being fundamental for relating observed composition to composition of cosmic rays in the sources.

Simulation studies of extensive air showers provide the basis for interpreting measurements of cascades induced by the highest-energy cosmic rays. Calculations of neutrino fluxes are important to establish the background for underground experiments, such as the search for nucleon decay, and to determine the level at which neutrino astronomy may be possible. Another subject of great current interest is the calculation of flux limits on magnetic monopoles from galactic magnetic fields and neutron star brightness.

Even though the computations sometimes require use of large computers, theoretical work in this field is inexpensive relative to the observational. Nevertheless, it is vitally important that it be nurtured and maintained.

18

Recommendations

SPACEBORNE EXPERIMENTS

We concur with the recommendations of the Astronomy Survey Committee for two moderate programs that are pertinent to spaceborne studies of cosmic rays (see G. B. Field, chairman, *Astronomy and Astrophysics for the 1980's*, Volume 1, National Academy Press, Washington, D.C., 1982). First, they recommended "an immediate and substantial augmentation to the NASA Explorer satellite program," and they went on to note that "among the scientific areas that at present appear to offer special promise for additional Explorer-class missions are the following, . . . A study of the isotopic and elemental composition of low-energy Galactic cosmic rays and solar energetic particles in the interplanetary medium." Also, that report said, "The Astronomy Survey Committee recommends a series of cosmic-ray experiments in space, to promote the study of solar and stellar activity, the interstellar medium, the origin of the elements, and violent solar and cosmic processes."

The report of the Cosmic-Ray Program Working Group (National Aeronautics and Space Administration, 1982) and the supplement to that report (National Aeronautics and Space Administration, 1985) outline a program that will achieve these objectives of the Astronomy Survey Committee and will take advantage of the opportunities described in Chapter 17 in the section on Spaceborne Experiments. *We recommend implementation of this program* as summarized below.

This recommended program includes two major new programs: (1) development of a Superconducting Magnetic Spectrometer Facility for the Space Station, which will permit "a series of cosmic-ray experiments" as suggested by the second Astronomy Survey Committee recommendation above, and (2) a Cosmic-Ray Composition Explorer that is essentially the Explorer described in the first Astronomy Survey Committee recommendation above. This program also includes other recommendations that are important for the vitality of cosmic-ray research.

We note that there are a few active research groups in other countries carrying on balloonborne and spaceflight cosmic-ray experiments. In the past there has been international cooperation, with complementary experiments from different countries on the same spacecraft or cooperative international development of a single experiment. We would expect this cooperation to continue in the future, particularly with the development of the Superconducting Magnetic Spectrometer Facility.

Major New Programs

As our highest priority, we recommend the development of a Superconducting Magnetic Spectrometer Facility for the Space Station capable of conducting a wide variety of measurements on the energetic galactic particles above 1 GeV. The heart of the facility would be a superconducting magnet and trajectory-defining detectors that would have a maximum detectable rigidity of several thousand GV. Above and below the magnet would be a variety of Cerenkov counters and energy-loss detectors, with the individual ancillary detectors being changed from time to time in order to optimize the detector configuration for various scientific objectives.

This magnet facility would permit a series of significant cosmic-ray observations. A search for antinuclei heavier than antiprotons would be possible with the unprecedented sensitivity of 10^{-8}; the detection of even a small flux of heavy antinuclei would have a profound influence on cosmology. The spectrum of antiprotons would be measured up to about 1000 GeV, giving important information about cosmic-ray confinement in the Galaxy and conceivably displaying the signature of exotic processes such as the annihilation of photinos. A significant contribution of this facility would be measurement of isotopic composition with excellent statistics and mass resolution over an energy range previously inaccessible to isotope resolution; these measurements would provide important signatures of the nucleosynthesis of cosmic rays and other matter and would give us radioactive clocks at high Lorentz factor for probing time scales of cosmic-ray acceleration and galactic confinement. The facility would permit measurements of electron and positron spectra to about 1000 GeV, providing unique clues concerning the distribution in the galaxy of sites of cosmic-ray acceleration. The excellent momentum resolution of the magnet facility would make possible the measurement of energy spectra of cosmic-ray nuclei over a very wide energy region, from a few GeV/amu to several hundred GeV/amu with unprecedented resolution, making possible a sensitive search for spectral or temporal changes that could carry the signature of individual sources of cosmic rays.

Also as a high priority we recommend an Explorer-class mission on a spacecraft outside the magnetosphere to carry high-resolution experiments to resolve the individual isotopes and elements of galactic cosmic rays, solar energetic particles, and anomalous cosmic rays in the energy region below 1 GeV/amu. Using established techniques, these experiments would have sufficient mass resolution and collecting power to determine the detailed isotopic composition and the energy spectra of all elements up through atomic number 30, with exploratory measurements of heavier nuclei.

This Explorer mission would provide a detailed comparison of the elemental and isotopic structure of solar matter (from solar energetic particles), local interstellar matter (which is believed to be the source of the anomalous cosmic rays), and more distant galactic matter (which is the source of the galactic cosmic rays), thereby adding new dimensions to studies of the nucleosynthesis and subsequent evolution of both galactic and solar-system matter. In addition it would allow particle injection and acceleration processes to be studied on scales ranging from in situ observations of interplanetary shock acceleration, to flare acceleration on the Sun, to cosmic-ray acceleration in the galaxy.

Continuing Programs

An essential prerequisite for the major new programs described above is the availability of frequent, relatively low-cost opportunities for exposing new instruments to space. High-altitude balloons have provided these opportunities for many years and are likely to continue to be the best way to test new detector concepts, make modest scientific advances, and educate graduate students. Similarly, if low-cost, relatively fast turn-around opportunities can be developed for attached instruments on the Space Shuttle, those will also prove valuable.

Continued tracking of the Pioneer and Voyager spacecraft and near-earth IMP-8 will provide otherwise unattainable information about the modulation of cosmic rays in the heliosphere. The cosmic-ray experiment on Ulysses (formerly called the International Solar Polar Mission) and the cosmic-ray experiment that has been selected for the WIND spacecraft in the International Solar Terrestrial Program will be valuable additions to this network and will make valuable advances in our knowledge of cosmic-ray isotopes.

The Cosmic Ray Nuclei Experiment, which was successfully flown on Spacelab-2 in August 1985, made important measurements of cosmic-ray composition to a few TeV. Its upper energy is principally limited by the low statistics imposed by its short (less than a week) exposure. We endorse the NASA decision to fly this experiment again on another Spacelab flight, and we strongly recommend placing this instrument on the Space Station for at least a year. With such an extended exposure it will be able to measure directly the cosmic-ray composition at energies where ground-based observations suggest a change of composition. Such a change is expected from some models of cosmic-ray acceleration, and measurement of the composition at these energies is important for testing these models.

We endorse the NASA decision to develop the Heavy Nuclei Collector, a very-large-area plastic-track detector to be launched in 1987 on the Long Duration Exposure Facility. This experiment will be capable of measuring actinide nuclei in the cosmic rays with high enough resolution and statistics to use these radioactive elements to measure the time scale since nucleosynthesis of the heavy cosmic rays.

Interpretation of measurements of cosmic-ray composition depends critically on knowledge of partial cross sections for spallation of heavy nuclei in collision with the interstellar gas. A continued program of measurement of such cross sections using the Bevalac heavy-ion accelerator is essential.

We recommend continued support of theoretical investigations related to particle astrophysics, including studies of shock acceleration and of the interrelated problems of injection-acceleration-confinement of cosmic rays.

Studies for the Future

A number of important measurements have been proposed in addition to those for which we have given high-priority recommendations above. Several of those deserve further study for possible implementation during the last few years of this century and the beginning of the next.

The Space Station will make possible assembly in space of very large instruments. We can identify three such devices whose feasibility should be studied: a high-energy array capable of measuring cosmic rays to 10^{16} eV, a large electronic detector capable of detecting hundreds of the rarest actinide nuclei and determining their energy spectra, and a spaceborne down-looking detector capable of observing the atmospheric scintillation from air showers of the highest-energy cosmic rays.

A study should be made of sending a new advanced set of instrumentation out of the heliosphere to measure a wide variety of interstellar parameters at distances of at least 100 AU.

Polar-orbiting platforms are part of the plans for the Space Station. Planning for these polar platforms should take into account the value of high-inclination orbits for studies of cosmic rays at moderate energy.

GROUND-BASED EXPERIMENTS

Ground-based experiments are supported by the the National Science Foundation (NSF) and the Department of Energy (DOE) and consequently have not received attention by NASA panels and working groups. Major NSF programs such as the Fly's Eye have been reviewed by the National Science Board as well as the normal referee procedures and the NSF physics advisory committee. The DOE-supported programs have been administered through the Division of High Energy Physics. In 1982-1983 the DOE convened an ad hoc advisory panel to advise it on experiments related to elementary-particle physics not using high-energy particle accelerators. All the DOE-supported programs have been reviewed by the Experimental Technical Assessment Panel (ETAP). The recommendations articulated here concur with the conclusions of ETAP and of the NSF advisory structure in every instance where the questions have been addressed. In response to increased activity in the field, DOE set up in late 1985 a standing High Energy Physics Advisory Panel (HEPAP) Subpanel on Non-Accelerator Particle Physics.

Most of the ground-based cosmic-ray experiments involve a group of physicists, an equipment inventory, and a budget on the scale of a typical experiment in particle physics in the external beam of a particle accelerator; and many are carried out by high-energy physicists. The total U.S. effort in the ground-based cosmic-ray experiments is much less than 1 percent of the particle-physics budget.

This contrasts with programs directed toward similar physics questions abroad. The Soviet Union and Japan spend relatively a much larger fraction of their effort here; and nations without high-energy particle accelerators, such as India, Australia, and Brazil, also have a relative commitment much greater than that of the United States. In spite of their modest scale, the U.S. programs remain internationally preeminent. It is appropriate that we continue to choose carefully the experimental efforts in this area and to support them vigorously. It may be noted that Japan, China, the Soviet Union, and South America have quite extensive programs involving emulsion chambers and calorimeters at mountain observatories. In particular, the Soviet Union is building a very ambitious mountain-top experiment in Armenia—the ANI. We do not recommend similar programs for the United States at this time.

Gamma-Ray Astronomy

The observation of gamma rays of energies above 10^{12} eV through ground-based observation of the Cerenkov light from air showers and direct detection of electrons from larger air showers provides the strongest evidence of discrete astronomical sources of acceleration processes extending beyond 10^{15} eV. Careful measurement of the direction and time structure of such showers has revealed several such sources, and the promise of further significant discoveries is very high. In order to exploit this recently developed field an expanded effort in utilizing existing detectors and in building new detectors is occurring.

- We recommend programs in gamma-ray astronomy as our highest priority ground-based cosmic-ray observation.

Highest-Energy Cosmic Rays and Extensive Air Showers

The Fly's Eye Program is unique among experiments around the world for studying cascades of 10^{18} eV and higher energies, and it serves as a focus for cosmic-ray research at the highest energies in the United States.

- We recommend also as a very high priority continued support of Fly's Eye and its improvements.
- We endorse studies of possible complementary surface detectors such as muon counters and scintillation-counter air-shower detectors.

These could expand the value of the Fly's Eye observations and could lead to new approaches to this energy region in the future.

High-Energy Neutrino Astronomy

As with gamma rays, neutrinos can provide in principle a line-of-sight signal from energetic astronomical sources of cosmic rays, penetrating regions that might be opaque to all electromagnetic radiation. Unfortunately, this very penetration is also related to the significant difficulty and cost in detecting such neutrinos. Neutrino detectors discussed, planned, and proposed include the MACRO detector in Italy and DUMAND of Hawaii. The full-scale DUMAND detector would be more expensive than any single ground-based experiment discussed here, and there are serious enough questions concerning that proposal to reserve judgment concerning its construction pending results from the prototype. At the same time, the discoveries that would result from a serious look at neutrinos of 10^{12} eV and above from astronomical sources would be exciting.

- We recommend that funding should be sought for neutrino astronomy detectors if their feasibility and cost-effectiveness can be clearly established.

Magnetic Monopoles

The search for these theoretically predicted, elusive, but fundamentally significant particles should be continued and extended.

- We suppport the construction of scintillation- or proportional-counter detectors capable of at least reaching the Parker bound, corresponding to at least a 1000-m^2 area.
- Larger flux-loop detectors of areas of the order of 100 m^2 should be built, and searches for monopoles trapped in meteorites or magnetite should be extended.

Large Underground Detectors

The upgrading and expanded exploitation of these detectors should be encouraged and supported. Justified and built to search for proton decay, these detectors are also valuable for cosmic-ray studies. These include the following:

- The study of neutrinos from the interaction of cosmic rays in the atmosphere and the search for neutrino oscillations as evidence for finite neutrino masses.
- The search for neutrino bursts from gravitational collapse of supernovae and other astronomical sources of neutrinos below 1 TeV.
- The study of muons underground, especially when coupled with surface air-shower arrays, in order to better understand primary composition in the 10^{13}-10^{16} eV energy region.

Solar Neutrinos

This problem, addressed by astronomers and particle physicists as well, merits continued serious effort. We support the following:

- Construction of detectors for neutrinos of lower energy through inverse beta decay, such as the proposed gallium experiment.
- Exploration of the feasibility of electronic detection of ν-e scattering in large underground detectors or other devices.

THEORY

The theoretical calculation and modeling of various processes are vitally important to continued progress in understanding cosmic-ray physics. The major theoretical activities concern the following:

- Stellar and explosive processes leading to generation of cosmic-ray nuclei and related photons and neutrinos.
- Acceleration mechanisms and propagation.
- Interactions and cascading of cosmic rays in the atmosphere and in the interstellar medium.

New concepts and the synthesis of ideas can lead to breakthroughs in our understanding quite out of proportion to the investment.

Index

A

Acceleration
  cosmic-ray, 128
    termination of, 130-131
  fractionation, 129, 130
  galactic, 129
  gravitation and, 15
  shock, 128-129
  solar, 128
Actinide elements, 126, 127
Active masses, 23
Adiabatic perturbations, 95
Advanced X-Ray Astrophysics Facility (AXAF), 101, 103-104
Air showers, 120
Air-shower
  detectors, 150
  experiment, 140-141
  observations, 130
Anisotropic cosmological models, 94
Anisotropy of space, 23
Antimatter, 148
Antinuclei, 138
Antiprotons, 116, 122, 136
Area Theorem, 64, 73
Astronomy
  gamma-ray, 149-150, 151, 162
  neutrino, 151-152, 163

Astronomy Survey Committee, 5, 157
Astrophysical properties of neutron stars and black holes, 75, 76
Astrophysics, vii-viii
Atomic time versus solar-system time, 21-22
AXAF (Advanced X-Ray Astrophysics Facility), 101, 103-104
Axion forces, 78
Axions, 22, 97-98

B

Backaction-evasion technique, 44
Bar detectors, 43-44
  observations with, 54
  sensitivity and bandwidth, 52, 54
Bar-type gravity-wave detector, 44, 45
Baryon density, 90-92, 97
Baryons, 90
Beryllium, 122, 124
Big bang, initial singularity of, 65
Big-bang
  models, 60, 87-89
  nucleosynthesis, 90-92
Binary
  pulsar, 14, 34-35, 39, 42-43
    discovery of, 47
  x-ray sources, 150

Black holes, 34-35, 36, 60
  astrophysical properties of, 75, 76
  colliding, 67, 68
  as invisible mass, 97
  quantum particle creation by, 64
  rotating, 63
  supermassive, 63
Black-hole
  binary, 77
  dynamics, four laws of, 64
  jets, 26-27
Blackbody
  curve, 93
  radiation, primordial, 89

C

Cerenkov light, 131
Clocks
  atomic versus gravitational, 22
  gravitation effect on rate of, 18
  hydrogen-maser, 18
Closure density of universe, 39, 100
COBE (Cosmic Background Explorer), 101, 103
Compact stars, systems of, 34-35
Computation, 58, 77
Computer technology, 110
Corkscrew jets, 27
Cosmic Background Explorer (COBE), 101, 103
Cosmic Censorship Hypothesis, 62, 64-65, 72
Cosmic Ray Explorer, 6
Cosmic Ray Nuclei Experiment, 160
Cosmic rays, vii, 111-164
  anomalous component, 132
  connection with gamma and radio astronomy, 137
  correlation between anisotropy and energy, 135
  energy dependence of escape from galaxy, 133-134
  energy spectra of, 118, 119
  galactic, 132-133
    extragalactic versus, 117
  ground-based experiments with, 7, 149-155, 161-164
  high-energy composition and spectra, 146-147
  high-energy nuclear and particle physics and, 137-142
  highlights, 121-142
  hydrogen in, 129
  interest in, 113-114
  isotope ratios and, 125
  large underground detectors for, 163-164
  major discoveries in, 121-123
  opportunities, 143-156
  origin of, 131
  overview, 115-120
  primary, 133
  principal recommendations in, 6-7
  recommendations, 157-164
  secondaries from light nuclei, 135-136
  secondary, 133
  solar, 131
  solar modulation of, 149
  solar-system material versus, 115-116
  space program in, 6-7, 143-149, 157-161
  theory of, 155-156, 164
Cosmic strings, decaying, 48
Cosmic-ray
  acceleration, 128
  cascades, 120
  detectors, nucleon decay experiments as, 138-139
  electrons, 135-136
  isotopes, galactic, 144-145
  lifetime, 136-137
  luminosity, 128
Cosmic-Ray Composition Explorer, 158, 159
Cosmological
  constant, 74, 92
  models
    alternative, 107
    anisotropic, 94
Cosmology, vii, 83-110
  features of, 85-86
  Grand Unification Theories and, 98-99
  ground-based studies in, 6, 104-106, 109
  highlights, 90-100
  opportunities, 101-107
  particle physics and, 106
  principal recommendations in, 5-6
  recommendations for, 108-110
  research, 6
  space program in, 5, 101-104

standard model of, 87-89
theory of, 106-107
CPT invariance, violation of, 78

D

Dark-matter problem, 86, 92, 96-98, 105-106
Deceleration parameter, 92, 103
Deep Underwater Muon and Neutrino Detector (DUMAND), 151, 163
Deuterium
  abundance of, 92
  observations of, 147
Deuterons, 90
Dipole effect, 94
Distance scale factor, 88
DUMAND (Deep Underwater Muon and Neutrino Detector), 151, 163

E

Earth-Moon distance, 17, 27, 30-31
Eclipse observations, 19
Einstein equations, 65-66, 77
  nonlinearities in, 66
Electromagnetic
  field, 12
  radiation, deflected, 19
  signal retardation, 19-21
Electrons, cosmic-ray, 135-136
Elementary particles, 88
Eötvös experiments, 15, 17, 22
Equivalence principle, 15, 31
Escape length, cosmic-ray, 133
ETAP (Experimental Technical Assessment Panel), 161
Euclidean functional integrals, 70
Expansion rate of universe, 103
Experimental Technical Assessment Panel (ETAP), 161
Extragalactic radio sources, 93

F

Field equation, 13
Fly's Eye detector, 7, 140-141, 150, 152, 162-163
Fractionation, acceleration, 129, 130
Frame-dragging precession, 24-25

G

Galactic
  acceleration, 129
  cosmic-ray isotopes, 144-145
  cosmic rays, 132-133
  formation, 89, 95
  nuclei, 86
  redshifts, 94-95
Galaxies, 86, 94
  angular distributions of, 94
  energy dependence of cosmic-ray escape from, 133-134
  large-scale clustering of, 105
  primeval, 102
Galileo mission, 55
Gamma Ray Observatory (GRO), 101
Gamma-ray
  astronomy, 149-150, 151, 162
  bursters, 61
Gamma rays, 122
  high-energy, 131
Gauge theories, 69
General relativity, 3
  experimental tests of
    highlights, 15-23
    introduction, 11-14
    opportunities, 24-35
  Lagrangian for, 73-74
  numerical techniques in, 67
  Theory of, 60, 78-79
Geodetic precession, 25
GP-B (Gravity Probe B) (Relativity Gyroscope Experiment), 24-26, 31, 80
Grand unification mass scale, 78
Grand Unification Theories (GUTs), 78
  cosmology and, 98-99
Gravitation, vii, 9-82; *see also* Gravity
  acceleration and, 15
  binding energy, 15, 17, 68
  clock versus atomic clock, 22
  collapse, 62
  constant, 92
    changing, 21
    rate of change of, 3
  effect on rate of clocks, 18
  effects, "magnetic," 24-27
  ground-based studies in, 4
  laboratory testing of, 22-23
  lenses, 100
  progress in study of, 11
  quadrupole moment of Sun, 33-34

recommendations in, 3-5, 80-82
redshift effect, 17-18
solar-system tests of theories of, 16
space program for, 4
theory, 5
  highlights, 61-71
  introduction, 59-60
  opportunities, 72-79
  recommendations, 80-82
Gravitational waves, 4, 36, 60
  detecting impulsive, 51
  detecting periodic, 52
  detecting stochastic, 53
  emission of, 67
  event rates and source calculations, 57
  search for
    highlights, 42-48
    introduction, 36-41
    opportunities, 49-58
  sources of, 38-40
    recent developments, 47-48
    spacecraft tracking and, 55-56
  theory of, 37
Gravitational-wave
  background noise, 39
  detectors, 40-41
Gravitino, 98
Gravitoelectric field, 12
Gravitomagnetic
  effects, 24-27
  field, 12
Graviton, nonlinear, 66
Gravity, 60; *see also* Gravitation
  alternative theories of, 65
  metric nature of, 3
  quantization of, 106
  quantum theory of, 59, 69-71, 73-75
Gravity Probe B (GP-B), 24-26, 31, 80
Gravity-wave detector, bar-type, 44, 45
GRO (Gamma Ray Observatory), 101
Ground-based studies
  in cosmic rays, 7, 149-155, 161-164
  in cosmology, 6
    continued, 104-106
    recommendations for, 109
  in gravitation, 4, 81
GUTs, *see* Grand Unification Theories
Gyroscopes, 24-26

H

H-space, 66
Hamiltonian of supergravity, 69
Hawking radiation, 59, 60, 64, 69
Heavy Nuclei Collector, 160
Helium, observations of, 147
HEPAP (High Energy Physics Advisory Panel), 161
Higgs
  fields, 99
  particles, 106
High Energy Physics Advisory Panel (HEPAP), 161
Hubble Space Telescope (HST), 101, 103, 104
Hubble's constant, 92, 100
Hubble's law, 94
Hughes-Drever experiments, 23
Hydrogen in cosmic rays, 129
Hydrogen-maser clock, 18

I

Impulsive gravitational waves, detecting, 51
Infinity, null, *see* Null infinity
Inflationary universe, 94, 99-100
Infrared Astronomy Satellite (IRAS), 101-102
Interferometric detectors, 44, 46
Invisible mass, 86, 92, 96-98, 105-106
IRAS (Infrared Astronomy Satellite), 101-102
Isotope
  abundances, 91
  ratios, cosmic rays and, 125
Isotopes, 144-145
  galactic cosmic-ray, 144-145
  solar-flare, 145
Isotropy, cosmological, 94

J

Jets, black-hole, 26-27

K

Kaluza-Klein theories, 71
Kerr solution to Einstein equations, 65

L

Lagrangian for general relativity, 73-74
Large Deployable Reflector (LDR), 101, 104
Large numbers hypothesis, 21
Laser interferometer detector, 40-41, 49-52, 56-57, 81

Laser ranging, *see* Range measurements
LDEF (Long Duration Exposure Facility), 144, 146
LDR (Large Deployable Reflector), 101, 104
Leaky-box model, 136-137
Light deflection by Sun, 19
Light-element abundances, 91
Long Duration Exposure Facility (LDEF), 144, 146
Long-Baseline Gravitational-Wave Facility, 4
Luminosity, cosmic-ray, 128
Lunar laser-ranging experiment, 17

M

MACRO (Monopole and Cosmic Ray Observatory), 152
"Magnetic" gravitational effects, 24-27
Magnetic monopoles, 99, 141-142, 153, 163
Mars Observer Mission, 29
Mass, missing, 86, 92, 96-98, 105-106
Mass-energy density, 87-88
Matter
  missing, 86, 92, 96-98, 105-106
  properties of, 85
Megaparsec (Mpc), 92
Mercury
  perihelion advance of, 21
  range measurements to, 28
Metric
  hypothesis, 12-13
  nature of gravity, 3
Microwave background radiation, 85, 87, 89
  absolute flux in, 93
  anisotropy in, 95, 96
Millisecond pulsars, 46-47
Missing matter, 86, 92, 96-98, 105-106
Monopole and Cosmic Ray Observatory (MACRO), 152
Monopoles, magnetic, 99, 141-142, 152-153, 163
Moon, range measurements to, 17, 27, 30-31
Mössbauer effect, 18
Mpc (megaparsec), 92
Muons, 154, 164

N

Naive quantum limit, 44, 54
Naked singularity, 62

National Aeronautics and Space Administration (NASA), 3-4, 24, 55, 80, 101-104, 108, 157-161
National New Technology Telescope, 5, 109
National Science Foundation (NSF), 4, 49, 109-110, 161
Neutrino
  astronomy, 151-152, 163
  mass, 97
  oscillation, 154
  types, 106
Neutrinos, solar, 123, 126-127, 154-155, 164
Neutron stars, 34-35, 39, 61-62
  astrophysical properties of, 75
  mass limit for, 92
  two, *see* Binary pulsar
Neutrons, 88
Nonlinear graviton, 66
NSF (National Science Foundation), 4
Nucleon, *see* Proton *entries*
Nucleosynthesis, 123-127
  big-bang, 90-92
Nucleus-nucleus interactions, 139-140, 148-149
Null
  experiments, 22
  infinity, 66
    angular momentum at, 73
    complex spaces at, 79

O

Orbital motion, 43

P

Parameter β, PPN, 31-32
Parameterized-post-Newtonian (PPN) formalism, 12-13
Particle Astrophysics Magnet Facility, 6
Particle physics, 71
  cosmology and, 106
Particle-antiparticle annihilation, 88
Particles, elementary, 88
Passive masses, 23
Periastron precession, 35
Perihelion advance of Mercury, 21
Periodic gravitational waves, detecting, 52
Perturbation theory, 69-70
Perturbations, adiabatic, 95
Photino, 98

Photon barrier, 89, 93
Physics
　cosmic-ray, see Cosmic rays
　gravitation, see Gravitation
　particle, see Particle physics
Planck era, 89, 99, 106
Planck mass scale, 78
POINTS (Precision Optical Interferometry in Space), 32-33, 81
Positive Energy Theorem, 68-69, 73-79
Positron-to-antiproton ratio, 147
Positrons, energy spectrum of, 136
PPN (parameterized-post-Newtonian) formalism, 12-13
Precession
　frame-dragging, 24-25
　geodetic, 25
　periastron, 35
　spin-orbit, 25
Precision Optical Interferometry in Space (POINTS), 32-33, 81
Preferred-frame effects, 13
Primary cosmic rays, 133
Primeval galaxies, 102
Primordial blackbody radiation, 89
Proton decay, 99, 106
　detectors, 153-154, 163
　　as cosmic-ray detectors, 138-139
　experiments, 138-139
Proton-proton cross section, 123
Protons, 88
Pulsar
　searches, 55
　timing, 46-47
Pulsars, 39
　binary, see Binary pulsar
　millisecond, 46-47
　radio, 46
　x-ray, 55

Q

Quadrupole
　anisotropy, 94
　moment, gravitational, of Sun, 33-34
　radiation, 43
Quantum
　effects in early universe, 64-65
　field theory in curved space-time, 69
　particle creation by black holes, 64
　theory of gravity, 59, 69-71, 73-75

Quantum-mechanical barrier penetration, 60
Quantum-nondemolition technique, 44
Quark-gluon
　phase, 140
　plasma, 148
Quarks, 88
Quasars, 63

R

Radar ranging, see Range measurements
Radio pulsar, 46
Radio sources, extragalactic, 93
Radio-interferometric techniques, 19
Range measurements
　to Mercury, 4, 28
　to Moon, 4, 17, 27, 30-31, 80
　solar-system, 4, 27-28, 80
　to Viking Landers on Mars, 28-29
Rapid r-process, 124, 125-126
Redshift effect, gravitational, 17-18
Redshifts, galaxy, 94-95
Relativity, 59
　general, see General relativity
　solar-system tests of, 77-78
Relativity Gyroscope Experiment (GP-B), 4, 24-26, 31, 80
Rotating
　black holes, 63
　stars, 62

S

Scalar-tensor theory, 11
Scale factor, distance, 88
Secondary cosmic rays, 133
Shock acceleration, 128-129
Shuttle Infrared Telescope Facility (SIRTF), 101, 102
Signal retardation, 19-21
Singularity
　initial, of big bang, 65
　naked, 62
　theorems, 59
SIRTF (Shuttle Infrared Telescope Facility), 101, 102
Slow s-process, 124, 125-126
Solar
　acceleration, 128
　corona, 121
　cosmic rays, 131

deflection of starlight, 32
flare isotopes, 145
modulation of cosmic rays, 149
neutrinos, 123, 126-127, 154-155, 164
quadrupole effect, 33-34
quadrupole moment, 29
system, 13
　formation of, 124
　measurements of dynamics of, 28
see also Sun *entries*
Solar-system
　material, cosmic rays versus, 115-116
　range measurements, 27-28
　tests, 4
　　of theories of gravitation, 16, 77-78
　　time versus atomic time, 21-22
Space
　anisotropy of, 23
　curvature, 12
　program
　　in cosmic rays, 6-7, 143-149, 157-161
　　in cosmology, 5, 101-104, 108-109
　　in gravitation, 4, 80-81
Space Shuttle, 113
Space Station, 113, 143-144, 161
Space-curvature effects, 19
Space-time, 85
　asymptotic properties of, 66-67, 72-73
　curved, quantum field theory in, 69
　foam, 74
　origin of, 107
　singularity, 62
Spacecraft tracking, gravitational waves and, 55-56
Spectral density, strain, 44
Spin-orbit precession, 25
Spin-spin coupling, 25
SQUIDs (superconducting quantum interference devices), 54
Starlight, solar deflection of, 32
STARPROBE, 31, 81
Stochastic gravitational waves, detecting, 53
Strain sensitivity, 39, 43, 54
Strain spectral density, 44
Sun, *see also* Solar *entries*
　light deflection by, 19
Sun-orbiting
　laser interferometer, 41, 56-57, 81
Superconducting
　coils, 153

quantum interference devices (SQUIDs), 54
Superconducting Magnetic Spectrometer Facility, 158-159
Superconducting Super Collider, 5
Supergravity, 71
　Hamiltonian of, 69
Supermassive black holes, 63
Supernova shock waves, 124
Supernovae, 38-39, 47, 103
Supersymmetric particle theories, 98

T

Termination of cosmic-ray acceleration, 130-131
Time-reversal invariance, violation of, 78
Twistor theory, 66

U

Ultraheavy elements, 145-146
Ulysses spacecraft, 55, 149
Uniqueness Theorems, 63
Universe(s)
　closure density of, 39, 100
　early, quantum effects in, 64-65
　expansion rate of, 103
　history of, 87, 88
　inflationary, 94, 99-100
　large-scale properties of, 92-94
　local, 86
　simple, 60
　structure in, 94-96
　uniqueness of, 85

V

Very Long Baseline Array, 5, 109
Viking Landers, 20
　range measurements to, 28-29

W

Weber bars, 40
Wormholes, 74

X

X-ray
　pulsars, 55
　sources, binary, 150